FIGURES IN LANGUAGE

DESCRIBE AND DRAW

A WORKBOOK OF COMMUNICATIVE ACTIVITIES

R. R. JORDAN

Collins ELT

By the same author

A Handbook for English Language Assistants (with R. Mackay, Collins 1976)
Listening Comprehension and Note-Taking Course (with K. James and A.J. Matthews, Collins 1979)
Reading in a Second Language (with R. Mackay and B. Barkman, Newbury House 1979)
Looking for Information (Longman 1980)
Academic Writing Course (Collins 1980)

1 2 3 4 5 6 7 8 9 10

Printed and published in Great Britain by
William Collins Sons and Co Ltd
Glasgow G4 ONB

ISBN 0 00 370006 2

This book is accompanied by a set of cassettes.

Contents

Acknowledgements

I am grateful to Teresa O'Brien for her assistance in recording material to go with this book, and to Terry McMylor for recording and editing the tapes. I am particularly grateful to Lucia Salamea for kindly providing feedback on the material in the book; and to Rosalind Grant-Robertson for a critical reading; I am also grateful to numerous overseas students at the University of Manchester for whom the material in this book was originally composed.

I wish to express my gratitude to Carol Whitlock and Pat Abdallah for typing the manuscript.

A word of thanks is due to the English-Speaking Union for organising their English Language Competition 1980. It provided the incentive for an outline version of this book which was awarded a prize. Acknowledgement is also made to the staff at the British Council's English Language Teaching Institute, London, for the original idea of 'Describe and Draw'.

I am grateful to the following for permission to make use of copyright material: *Facts in Focus*, Penguin Books, 4th edition 1978, for the data on which the following diagrams are based – U.K.: Some Statistics; United Kingdom: Population; G.B. Road Accidents: Number Killed; United Kingdom energy consumption. 'Recent Trends in Monopoly in Great Britain' by G. Walshe in *Occasional Paper 27*, 1974, by the National Institute of Economic and Social Research for the basis of the figure 'The Trend to Monopoly in British Manufacturing'. Source unknown for the diagram 'Factors Affecting Health in a Developing Country.'

Introduction

AIM OF THE BOOK

Background

Experience has shown that students who learn English as a foreign language usually have difficulty when they try to handle *non-verbal data* (e.g. facts and figures, tables and graphs) in the *spoken form* in English. They usually have difficulty in *understanding quickly* data which is presented as part of a talk or lecture in English and which is not accompanied by visual aids. They also have difficulty in *asking questions* in English about data and, perhaps, even more difficulty in *commenting* upon data or *describing* parts of it in a talk or in a discussion. In other words, there is a problem of *oral numeracy*. '*Figures in Language*' aims to help students overcome the above difficulties.

Purpose

This book is accompanied by a cassette which is an essential part of the material. It is intended to give oral practice in facts and figures by means of a technique known as '*Describe and Draw*' (this is explained in detail below). By working through the material individually, in pairs, or groups, or even a large class, students can improve their concentration in listening and can focus on details of the language which are important for accuracy.

The material can be used at language levels from elementary to advanced by varying the degree of detail which is acceptable in the figures. General numerical or mathematical language is used: this provides the basis for more specialised vocabulary (e.g. statistical or technical) which can be supplied by subject-specialist teachers later as it is needed.

HOW TO USE THE BOOK

The book is divided into two main parts. *Part 1 (A): Aids to Description* is a reference section that should be looked at before using the book. It can also be looked at while Part 2 is being used. It consists of language structures and vocabulary that will be of use when handling data. Letters and abbreviations are recorded on cassette. *Part 1 (B)* provides practice for some of the information in Part 1 (A). The material for it is meant to be read for individual use, but it can also be used in pairs or groups.

Part 2: Figures for Description contains a number of diagrams that are provided for students to describe to each other and to draw. The sub-sections are *Shapes* (6 increasingly complex shapes), *Data* (12 pages of various types of data display – tables, graphs, histogram, diagrams, etc.), *Maps and Plans* (6 pages of various types of maps and plans). Next to each figure is a *tapescript* of the description as a native English-speaker might describe it. *These descriptions appear on the accompanying cassette.* It is not necessary to be artistic in any way when drawing the diagrams. The aim is to reproduce the diagram, etc. as accurately as possible and, in the process, note how the language is used. *A pencil, a rubber and some paper will be needed.*

Method

STUDENTS

1. If you are using the book on your own: first read carefully Part 1 (A) and then do the exercises in Part 1 (B). In order to draw the figures in Part 2 listen to the cassette. Listen to the full description first on the cassette and then play it again in order for you to draw it. Stop the cassette when you wish to give yourself time to draw the parts of the diagram. Do *not* look at the figures in the book until you have finished drawing them from the description on cassette. When you have finished your drawing compare it with the illustration in the book. If you have difficulty, read carefully the corresponding tapescript and, if necessary, listen again to the cassette.

2. If you are using the book in pairs: do Part 1 (A) and (B) as above – or one person reads out the description in 1 (B) and the other draws it. Answers can be checked at the back of the book.

Part 2 should be used as follows: you, as student A, should open the book at the first illustration. Do *not* read the tapescript. You should first give student B a *very brief general description* of the picture as an introduction (perhaps two or three sentences). Then you should tell student B to draw exactly what you describe. *Do not watch him and do not let him see the diagram in the book.* You then describe the diagram slowly and carefully in detail, pausing to give time for student B to draw it. If you have difficulty in describing it you can refer to Part 1 (A). If necessary, you can look at the tapescript opposite the diagram. Tell student B that he may ask you questions at any stage about the diagram or your description in order for him to draw it accurately.

When student B has finished drawing show him the diagram in the book. Compare the two versions and discuss any difficulties.

You can then change with student B and he can describe the next diagram for you to draw. *It is important, therefore, not to look in advance at the subsequent diagrams.*

TEACHER

3. If you are the teacher and using the book with a group or class: first ensure that the class understands Part 1 (A). Then practise with them Part 1 (B): they can do this silently, reading the instructions; or you can read out the instructions to them. From Part 2 select one of the early diagrams for description – either describe it to the class yourself, or play the cassette of the description. Remember to allow time for drawing and for the class to ask you questions. If possible, let the students work in pairs after your initial demonstration. Your description need not, of course, match exactly the description given in the tapescript as there are several ways of describing each diagram.

4. Extensions

(1) After some of the diagrams and tables have been described, you could compose your own containing information of interest to your students. In this way they will obtain more practice in the language presented in this book.

(2) Some of the tables and diagrams can be used for detailed questions and answers after they have been completed. e.g. '*Some Numbers*': which number is in the top right-hand corner? '*Two Families*': When was June Davies born? When did she die? '*United Kingdom: Population*': What was the population of Scotland in 1921? '*A Timetable*': What have you got at 10.30 on Thursday? '*A page from a diary*': Where will you be at 11.20 a.m. on Tuesday?

5. Precision

You can decide how precise you wish to be in describing the diagrams or drawing the figures. You can describe with very rough measurements, or you can be exact. You can indicate whether or not capital or small letters should be used, and whether there should be underlining or not. You can also spell out difficult words.

6. General Recommendations

You will find it helpful to use a good English – English dictionary e.g. *The Advanced Learner's Dictionary* (OUP), *Longmans Dictionary of Contemporary English* (Longman), *Collins English Learner's Dictionary* (Collins). You may also find it useful to refer to the appendices at the back of these dictionaries for more information on mathematical and numerical expressions.

Finally, REMEMBER – do NOT look at the diagrams until after you have tried to draw them.

Part 1 (A) Aids to Description: Reference

Letters and Abbreviations

(1) **The English Alphabet:** Can you say these letters correctly? Listen to them on the cassette.

CAPITAL LETTERS: A B C D E F G H I J K L M N O P Q R S T U V W X Y Z

small letters:
(or lower case letters) a b c d e f g h i j k l m n o p q r s t u v w x y z

(2) **Abbreviations:** These are quite commonly used in everyday or academic life. Can you say them correctly? Listen to the cassette and try to write them down. For their meaning look in a good English dictionary.

A.A.	H.Q.	R.A.F.
A.D., B.C.	I.B.M.	R.I.P.
A.G.M.	I.C.I.	R.S.A.
a.m., p.m.	i.e.	R.S.P.C.A.
B.A., B.Sc., M.A., M.Sc., Ph.D.	I.O.U.	R.S.V.P.
B.B.C., N.B.C., C.B.C., etc.	I.R.A.	S.O.S.
c.c.	I.T.V.	S.R.N.
C.I.A.	J.P.	S.T.D.
C.I.D.	K.G.B.	T.B.
C.O.D.	L.P.	T.U.C.
C.S.E., G.C.E.	MI5	T.V.
D.D.T.	M.P.	U.H.F., V.H.F.
D.J.	m.p.h./k.p.h.	U.K.
D.N.A., R.N.A.	N.B.	U.N.
E.E.C.	N.H.S.	U.S.A.
e.g.	N.S.P.C.C.	U.S.S.R.
F.A.O., I.L.O., I.M.F., W.H.O.	O.K.	V.C.
F.B.I.	P.A.Y.E.	V.D.
G.B.	P.C.	V.I.P.
G.M.T., B.S.T.	P.E.	V.S.O.
G.N.P.	P.M.	w.c.
G.P.	P.R.O.	Y.H.A.
G.P.O.	P.S.	Y.M.C.A.
H.M.S.O., O.H.M.S.	P.T.	Y.W.C.A.
h.p., H.P.	P.T.O.	

NOTE: Letters used in the abbreviations above are always spelled separately e.g. c.c. = spoken/SI SI/. When the letters are 'doubled' in words, the word 'double' is often used when spelling, e.g. accommodation is spelled a- *double* c -o- *double* m -o-d-a-t-i-o-n. (The term 'double' can sometimes be used in giving phone numbers, e.g. 2744 = two seven double four.)

Numbers, Dates and Times

Cardinal Numbers		Ordinal Numbers	
1	one	1st	first
2	two	2nd	second
3	three	3rd	third
4	four	4th	fourth
5	five	5th	fifth
6	six	6th	sixth
7	seven	7th	seventh
8	eight	8th	eighth
9	nine	9th	ninth
10	ten	10th	tenth
11	eleven	11th	eleventh
12	twelve	12th	twelfth
13	thirteen	13th	thirteenth
14	fourteen	14th	fourteenth
15	fifteen	15th	fifteenth
16	sixteen	16th	sixteenth
17	seventeen	17th	seventeenth
18	eighteen	18th	eighteenth
19	nineteen	19th	nineteenth
20	twenty	20th	twentieth
21	twenty-one	21st	twenty-first
22	twenty-two	22nd	twenty-second
23	twenty-three	23rd	twenty-third
24	twenty-four	24th	twenty-fourth
30	thirty	30th	thirtieth
40	forty	40th	fortieth
50	fifty	50th	fiftieth
60	sixty	60th	sixtieth
70	seventy	70th	seventieth
80	eighty	80th	eightieth
90	ninety	90th	ninetieth
100	a/one hundred	100th	a/one hundredth
1000	a/one thousand	1000th	a/one thousandth

0 = nought, zero: in mathematics, decimals
0 = oh: in telephone numbers, dates, times
0 = nil: in many sports and games
0 = nothing: general

Note how the following are said:

108 a hundred *and* eight
4,006 four thousand *and* six
4,105 four thousand, *one* hundred *and* five

Dates: Some Years

1900	nineteen hundred
1907	nineteen oh seven
1919	nineteen nineteen
1923	nineteen twenty-three
1970	nineteen seventy
1990	nineteen ninety
2000	two thousand

Dates: Some Days and Months

. . . written . . .	*. . . spoken . . .*
Wednesday 13th July	(the 13th of July)
Thursday 30th November	(the 30th of November)
Tuesday 22nd February	(the 22nd of February)
Friday 23rd August	(the 23rd of August)
Saturday 31st May	(the 31st of May)

Note the use of 1st, 2nd, 3rd but 4th, 5th etc.

Some Times

(a.m. = morning;
p.m. = afternoon/evening)

8 (a.m. or p.m.)	eight o'clock
6.05	five past six (or: six five)
7.10	ten past seven (or: seven ten)
9.15	a quarter past nine (or: nine fifteen)
3.20	twenty past three (or: three twenty)
11.25	twenty-five past eleven (or: eleven twenty-five)
4.30	half past four (or: four thirty)
2.35	twenty-five to three (or: two thirty-five)
1.40	twenty to two (or: one forty)
5.45	a quarter to six (or: five forty-five)
11.50	ten to twelve (or: eleven fifty)
10.55	five to eleven (or: ten fifty-five)

Measurement, Direction and Location

(1) 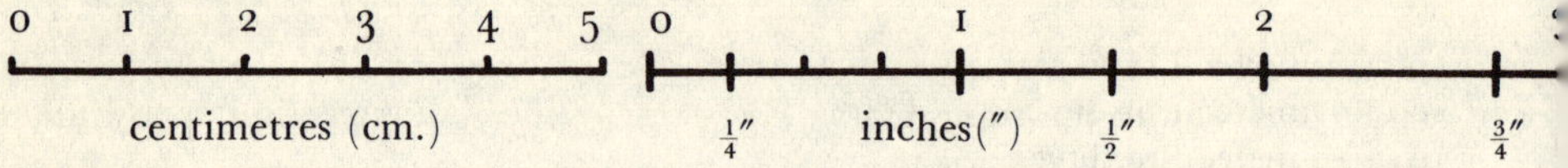

Fractions

$1\frac{1}{4}$ one and a quarter
$2\frac{1}{2}$ two and a half
$3\frac{2}{3}$ three and two-thirds
$4\frac{3}{4}$ four and three-quarters

($\frac{1}{4}''$)	a quarter of	an inch
($\frac{1}{2}''$)	half	
($\frac{3}{4}''$)	three-quarters of	
($\frac{1}{3}''$)	a third of	
($\frac{2}{3}''$)	two-thirds of	

Decimals

.1 point one
0.6 nought point six
0.07 nought point nought seven
1.008 one point nought nought eight
2.75 two point seven five
32,400.25 thirty-two thousand, four hundred point two five.

Distance

11 in. eleven inches
27 cm. twenty-seven centimetres
85 m. eighty-five metres
104 km. a hundred *and* four kilometres
253 miles. two hundred *and* fifty-three miles

---

(2) **Symbols**

+ plus
− minus
× times, multiplied by, by
÷ divided by
± plus or minus
= equals, is equal to
% percent(age)
& and

Punctuation

, comma
. full stop
: colon
; semi-colon
? question mark
! exclamation mark
“ ” inverted commas (quotation marks)
- hyphen (when joining parts of a word)
( ) brackets
’ apostrophe
– dash
/ stroke

---

(3) **Powers (Indices) and Roots**

3^2 three *squared*
5^3 five *cubed*
4^5 four *to the power of* five
2^7 two *to the power of* seven

$\sqrt{9}$ the square root of nine
$\sqrt[3]{8}$ the cube root of eight
$\sqrt[5]{x}$ the fifth root of x
$\sqrt[7]{x}$ the seventh root of x

Equations often make use of brackets ( ), [square brackets] and {curly brackets}

---

(4) **Scale**

on maps: 1:25,000 one to twenty-five thousand (or $2\frac{1}{2}$ inches to 1 mile)
1:50,000 one to fifty thousand (or 2 cm. to 1 km.)
1:63,360 one to sixty-three thousand three hundred and sixty (or 1 inch to 1 mile)
the scale of the map is 5 miles to the inch

on plans: 1:60 the plan is drawn to a scale of one to sixty

(5)

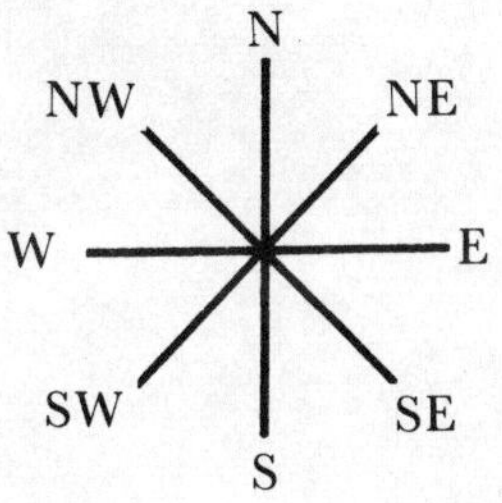

Compass points . . . adjectives

north	(N)	northern
north-east	(NE)	north-eastern
east	(E)	eastern
south-east	(SE)	south-eastern
south	(S)	southern
south-west	(SW)	south-western
west	(W)	western
north-west	(NW)	north-western

X is	to	the	north east	of the	capital
	in		south west etc.		country etc.

e.g. X is to the east of the capital
X is in the south of the country

---

(6)

to	the	right
on		left

on the	right-hand	side
in the	left-hand	column

in the	top bottom	left-hand right-hand	corner

at the	top bottom side

in the	middle centre corner

the		column	on	the	right
the	second third fourth etc.		from		left

e.g. the third column from the left

the	second third fourth fifth etc.	row line	up down	
			from the	top bottom

e.g. the fourth row from the top

(7) **Points of reference in a table**

top COLUMNS

the top left-hand box; or: the left-hand column, top row	the middle (or centre) column, top row	the top right-hand box; or: the right-hand column, top row	ROWS (LINES)
the left-hand column, second row (down)	the middle column, second row (down)	the right-hand column, second row (down)	
the left-hand column, third row (down)	the middle column, third row (down)	the right-hand column, third row (down)	
the bottom left-hand box; or: the left-hand column, bottom row	the middle column, bottom row	the bottom right-hand box; or: the right-hand column, bottom row	

left *right*

middle (or: centre)

bottom

In the above table, there are three columns and four rows.

(8) (i) **Vocabulary**

Some useful words for giving and carrying out instructions for describing and drawing (examples of their use will be found in the tapescripts). *If you do not know the words, look up their meaning in a dictionary before you begin.*

NOUNS: drawing, shape, design, diagram, figure, table, graph, histogram/bar diagram, map, plan;
data, axis/axes, cross-section; star/asterisk, footnote.

VERBS: draw, listen, measure, spell, underline, write;
add, complete, contain, continue, extend, finish, form, include, represent, start, stop;
indicate, label, mark, plot, point, shade;
cross, divide (sub-divide), join, link, meet, touch.

N.B. (1) Often it will be necessary to compare the size and position of items. In order to do this you need to be familiar with the comparative forms of many of the adjectives given in Part 1 (A), e.g. high – higher; wide – wider, etc.
(2) Similarly you may find it useful to use verbs formed from some of the adjectives, e.g. to lengthen = to make longer; to shorten = to make shorter.

(ii) **Pairs of prepositions and adjectives**

It is important for you to be able to distinguish quickly between the following pairs of prepositions and adjectives when using this book (this facility will be generally useful to you):

up	upper	over	above	top	left	horizontal	east
down	lower	under	below	bottom	right	vertical	west
		beneath					

(iii) **Pronunciation**

(1) Remember, '*east*' rhymes with 'least' or 'feast'
'*west*' rhymes with 'best' or 'nest' } these are often confused.

(2) In the '*Two Families*' tables boys' and girls' names are used: these are not always easy to say:

girls: '*Jean*' rhymes with 'seen' or 'bean'.
'*June*' rhymes with 'moon' or 'spoon'.
'*Joan*' rhymes with 'bone' or 'phone'.
'*Jane*' rhymes with 'cane' or 'lane'.
boy: '*John*' rhymes with 'gone' or 'Ron'.

(3) Particular care needs to be taken with the pronunciation of the ends of some numbers, e.g. nine<u>teen</u> (19) and nine<u>ty</u> (90).

Shapes, Lines, Dimensions and Angles

Shapes
(& adjectives)

square
(square)

rectangle
(rectangular)

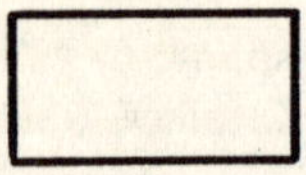

triangle
(triangular)

circle (circular)

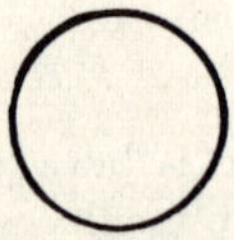

semi-circle
(semi-circular)

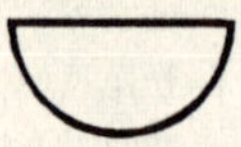

ellipse/oval
(elliptical)

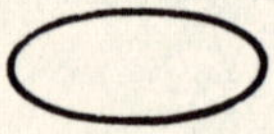

parallelogram

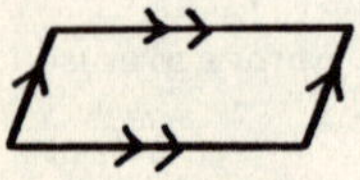

cube
(cubic)

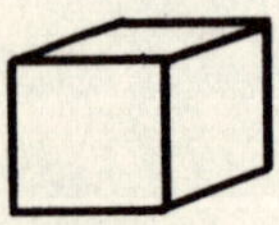

(convex)

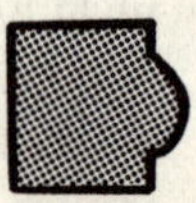

(concave)

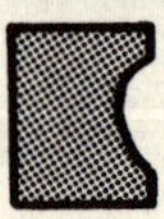

Lines – *all adjectives*

horizontal

vertical

diagonal
(also: sloping)

straight

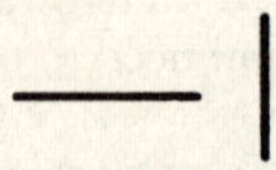

curved

wavy

parallel

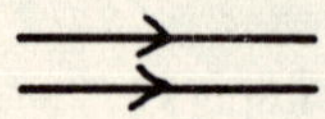

perpendicular:
AB is
perpendicular
to CD

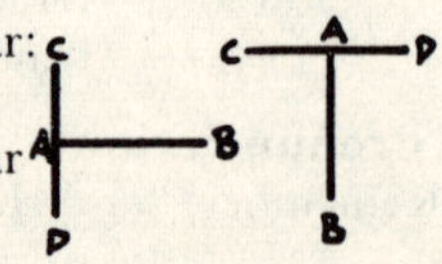

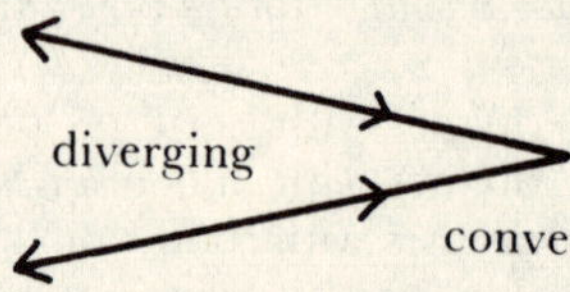

diverging

converging

solid

broken

(dashes)

dotted

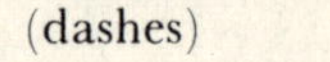

(dots)

Angles: *A triangle*

a right angled triangle

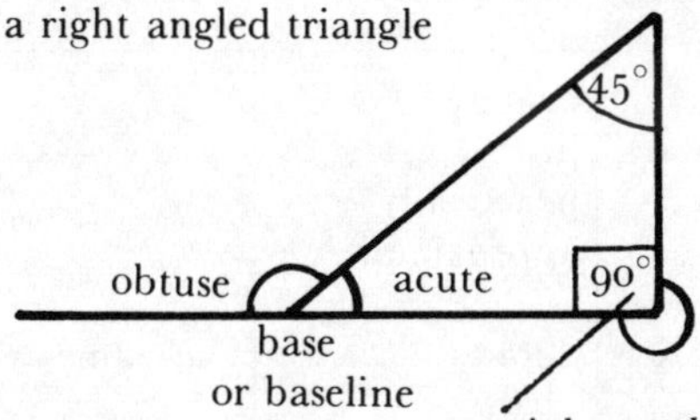

an equilateral triangle

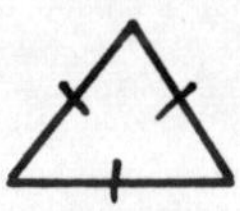

an isosceles triangle

Dimensions
(& adjectives)

length (long)
width (wide)
breadth (broad)
height (high)
depth (deep)
thickness (thick)
narrowness (narrow)
capacity
area
volume
circumference
diameter
radius

Dimensions

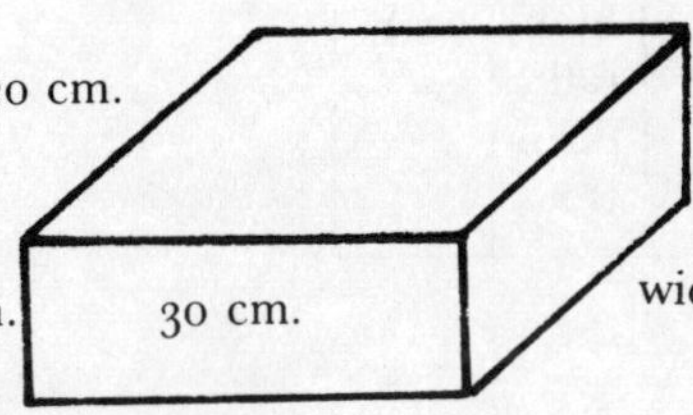

The *volume* is 6,000 c.c.
(cubic centimetres)

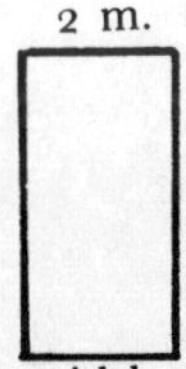

The *area* is
8 square metres (8 m^2)

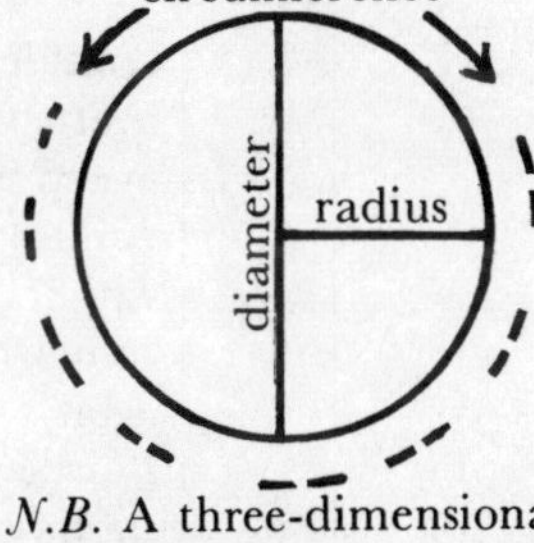

N.B. A three-dimensional circle is a *sphere*

an arc

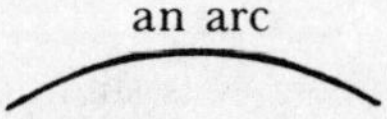

an arrow

Note:
(1) The above dimensions are for *solid* objects, e.g. bricks.

For an *open, hollow* object, e.g. a swimming pool, *depth* is used not height.

(2) 'Tall' is used instead of 'high' for people and some objects, e.g. trees.

Precision, Approximation, and Qualification of Comparison

Terms commonly used, listed in a rough order of exactness.

X is	exactly		5 cm.	long
	approximately		etc.	wide
	about			high
	around			deep
	more or less			thick
	roughly			
	a fraction			
	slightly	over		
	a little	under		
	just			

e.g. X is about 5 cm. wide.
It is just over 10 cm. long.

	exactly	
	just	
	practically	
	approximately	
X is	about	the same as Y
	more or less	
	almost	
	nearly	
	roughly	

e.g. X is nearly the same as Y.

	not (quite) as/so		big	
			long	
	half		wide	
X is	twice		high	as Y
	three times	as	tall	
	four times		deep	
	etc.		thick	

e.g. X is not as big as Y.
X is twice as wide as Y.

	considerably	smaller	
	a great deal	bigger	
	(very) much	longer	
	(quite) a lot	shorter	
X is	rather	wider	than Y
	somewhat	narrower	
	a bit	higher	
	a little	taller	
	slightly	deeper	
		thicker	
		thinner	

e.g. X is slightly thicker than Y.

Asking Questions for Information

(1) Notice how the following questions are formed with a *question-word*:

What is the length of the bottom line?
Where is the letter 'p'?
Where does the circle go?
Which corner does the line AB go to?
When was John Davis born?
Who (did you say) was born in 1717?
Whose class is at 11.30 on Thursday?

How long is the horizontal line?
How far should I draw the line?
How do you spell that please?
and: *How* big is it?
How long is it?
How wide is it?
How tall is it? etc.

(2) (a) Notice how the following questions are formed with an *auxiliary verb*:
Is the length 7 metres?
Are the two squares the same size?
Does the diagonal line slope to the right?
Do the two lines touch?

(b) They can also be used with *comparisons*:
Is the vertical line *longer than* the horizontal line?

(c) and with *alternatives*:
Is the length 17 metres *or* 70 metres?
Did you say 17 *or* 70?

Is it a	vertical	or	horizontal	line	?
	solid		broken		
	concave		convex	curve	

(3) *Requests for Repetition or Clarification:* frequently –
What was that you said? Could you say that again please? – but also as below:

Please could you	repeat	the	(last)	word	(?)	more	clearly	?
		that (?)		sentence			loudly	
				title			slowly	
				name				
				date				
				number				
				distance				
				quantity				
				measurement				
				etc.				
	spell			word	?			
				title				
				name				
				etc.				

e.g. Please could you repeat the last sentence more slowly?
Please could you spell that word?

Part 1 (B) Aids to Description: Practice

In this Part you can practise some of the figures and language used in Part 1 (A). If necessary, look again at Part 1 (A). The purpose of the practice is to help you with Part 2.

Types of practice: There are two ways that you can use most of the practice exercises given below:

(1) *read them to yourself* and carry out the instructions given, drawing the appropriate diagrams etc.

(2) *read them aloud* to a partner and he/she should draw/write according to your instructions.

The *answers* to these exercises are given at the back of the book.

Exercise 1

H	B	E
		G
A		
		C
	F	
D		

(1) Describe the location of the capital letters in the table. (e.g. H is in the top left-hand box.)

(2) Now put the following small letters in the boxes referred to in the table:

1 Write an 'i' in the left-hand column, fifth row down.
2 Write a 'j' in the middle column, fourth row down.
3 Write a 'k' in the left-hand column, second row down.
4 Write an 'l' in the middle column, bottom row.
5 Write an 'm' in the right-hand column, third row down.

(3) Decide in which boxes to put the letters, *n, o, p, q, r*; then describe to another student where they are. The other student should write the letters in the correct boxes. Check his/her answers when you have finished.

Exercise 2

Draw a horizontal line about 5 cm. long. At the left-hand end of the line write the capital letter A. At the right-hand end write B. Now draw another line of the same length which crosses AB vertically in the centre. The vertical line is *bisected** by AB. *Label** the vertical line CD: C is above AB, and D is below.

NOTE: **to bisect* = to cut or divide into two (usually equal parts)
**to label* = to write on something what it is

Exercise 3

Draw the vertical line PQ (any length). It is crossed by the horizontal line RS (also any length). Lines PQ and RS *intersect** at point X.

NOTE: **to intersect* = (of lines) to cut or cross each other

Exercise 4

Draw a square whose sides measure about 4 cm. Write the figure 4 inside the square, next to the left-hand vertical line and in the centre. Under the bottom line, to the right, draw a very small circle which just touches it.

Exercise 5
Draw a small rectangle with the longer side horizontal. Label the bottom left-hand corner A outside the rectangle, and the top right-hand corner B also outside. Draw a diagonal line joining AB. Label the bottom right-hand corner C, also outside the rectangle.

Exercise 6
Draw a rectangle, measuring roughly 6 cm. by 3 cm., the long side lying horizontal. Draw two lines, one dividing the rectangle vertically into two halves, the other dividing the box horizontally in half. In the top left-hand box write the word 'year'; in the box next to it on the right, write 'output'. In the box immediately below 'year' write, in figures, '1917'. In the box next to that write, also in figures '23,067 tons'.

Exercise 7
Give instructions for drawing the following diagram.

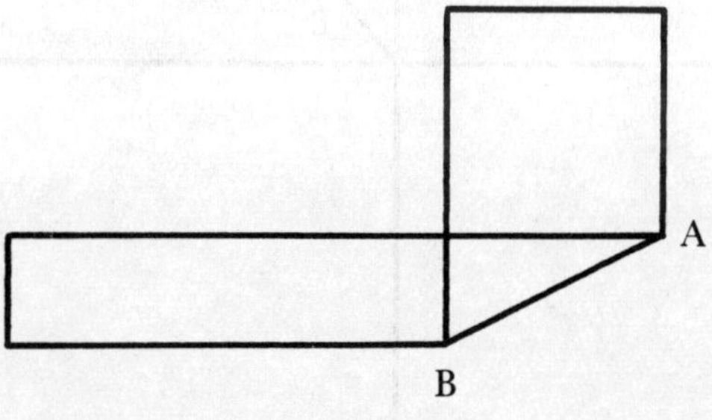

Different Shapes

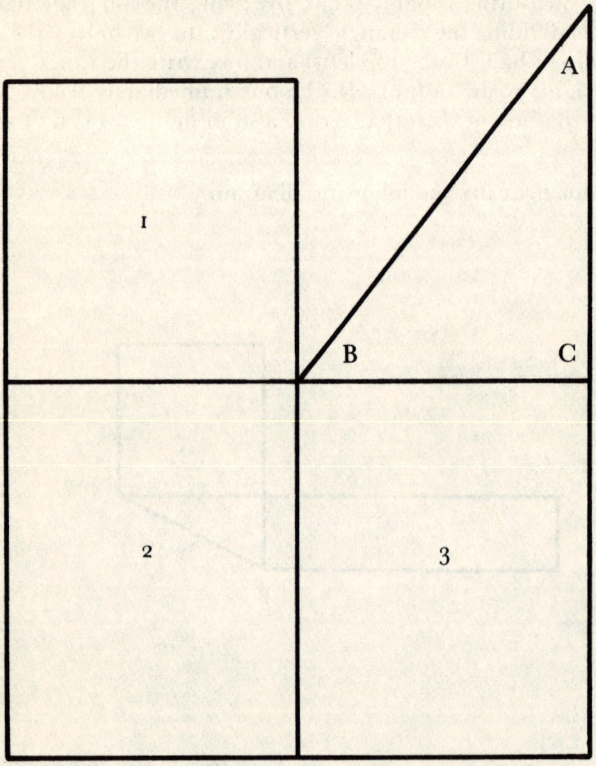

Part 2 Figures for Description

1 Shapes

Different Shapes

Introduction. The diagram consists of four shapes which are joined together. The size of the shapes does not matter.

Description. First, near the bottom of the page, and in the centre, draw a rectangle that is lying horizontally, measuring about eight centimetres by five. Draw a vertical line down the centre of the rectangle so that it is divided into two halves. You now have two smaller rectangles. Directly on top of the left-hand rectangle draw a square whose sides are the same length as the top of the left-hand rectangle. When you have finished that, write the number one in the centre of the square. Then write number two in the centre of the rectangle immediately below the square. In the rectangle to the right of that, put number three, also in the centre.

The final part to draw is on top of box number three. Listen carefully to the following description before drawing it. Draw a triangle so that it stands on top of the right-hand rectangle. The length of the base is the same as the side of the rectangle. Draw the right-hand side of the triangle as a continuation of the right-hand side of the rectangle. Draw it vertically in a straight line so that it goes about a centimetre higher than the square. Now join the top of that line to the top left-hand corner of box three.

When you have finished that, write capital letters inside the corners of the triangle: put A in the top corner, B in the bottom left corner, and C in the bottom right corner.

A Large Square Containing Four Shapes

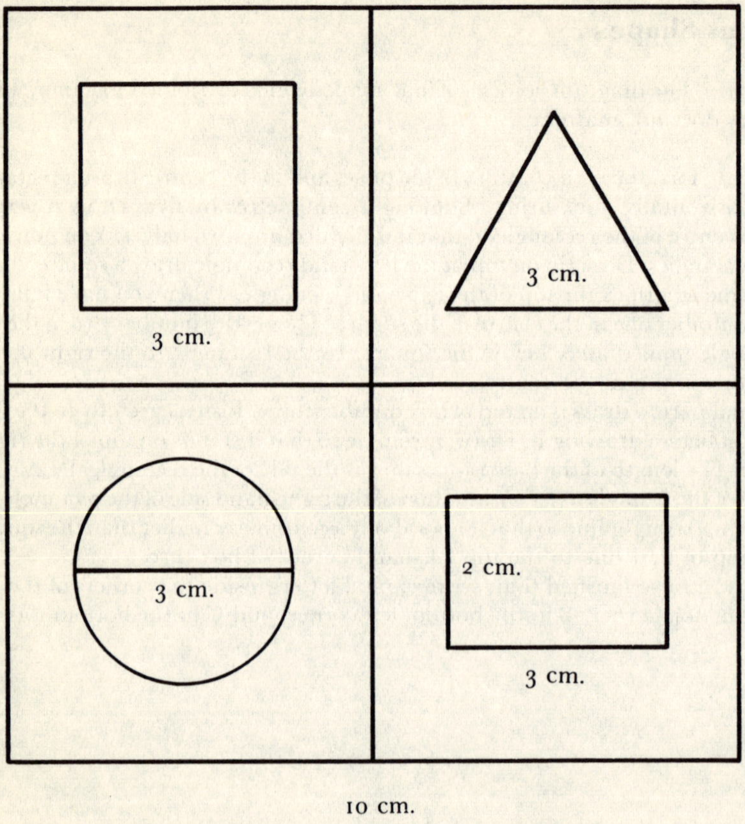

A Large Square Containing Four Shapes

Introduction. This is a drawing of a large square containing four smaller squares. Inside each of the four squares there is a different shape.

Description. In the middle of your paper draw a large square, measuring about ten centimetres. Write, in figures, 'ten cm.' under the baseline of the square and in the middle. Now draw a vertical line from the centre of the top side of the square to the centre of the baseline of the square. Draw a horizontal line from the centre of the left-hand side of the square to the centre of the right-hand side. Now you will have a large square and inside it there will be four smaller squares.

Inside the top left-hand square, and in the centre of it, draw another, smaller square, measuring about three centimetres. Write the figure three and the abbreviation 'cm.' just under the baseline of the smaller square. In the next square, to the right of the first square, draw an equilateral triangle in the middle. Each side measures about three centimetres. Write 'three cm.' inside the triangle and just above the baseline.

Inside the bottom left-hand square, in the centre, draw a circle whose diameter is three centimetres. Draw a horizontal diameter line. Write 'three cm.' inside the circle just below the diameter line. Now look at the bottom right-hand square. Inside it, in the centre, draw a rectangle whose horizontal length is about three centimetres and its width is two centimetres. Write 'three cm.' just below the baseline of the rectangle and 'two cm.' just inside the rectangle next to the left-hand vertical line.

A Two-dimensional Figure

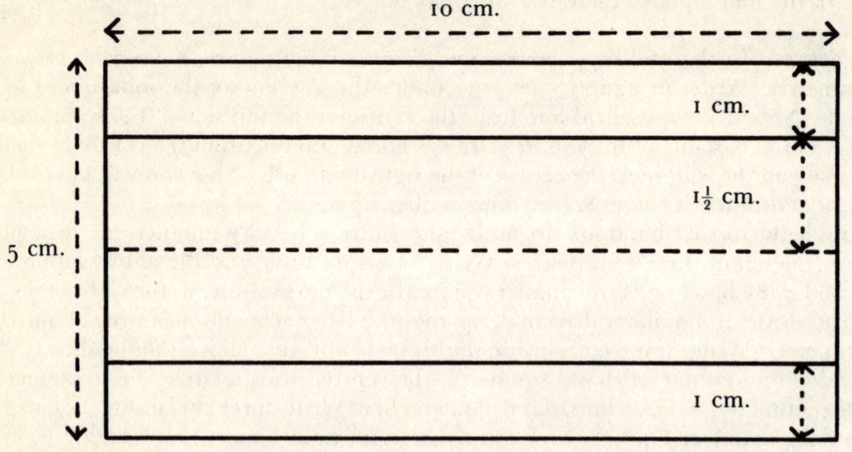

A Two-dimensional Figure

Introduction. The diagram is a large rectangle containing some parallel lines, with dimensions indicated next to arrowed broken lines.

Description. In the middle of your paper draw a rectangle lying horizontally. The length of the horizontal side is ten centimetres and the short side is five centimetres. Draw just above the top horizontal line, and parallel with it, a broken line which extends the whole length of the side. At each end of this broken line draw an arrowhead, and in the centre of the broken line, just above it, write in figures 'ten cm.'

To the left of the rectangle, just outside it, draw another broken line, parallel with the vertical side and extending the whole length of the side. Draw an arrowhead at each end, and in the centre of the line, to the left of it, write in figures 'five cm.'

Measure halfway down the vertical side of the rectangle, and then draw a horizontal broken line which divides the rectangle into two halves. One centimetre from the top of the rectangle, and inside it, draw a solid horizontal line parallel with the top line. Do the same at the bottom of the rectangle: in other words, one centimetre from the bottom line, inside the rectangle, draw a solid line parallel with the bottom line.

Inside the rectangle, on the right, next to the vertical line, draw vertical arrowed broken lines between the horizontal lines as follows. Firstly, in the top right-hand corner, between the top two solid lines, draw a vertical broken line with an arrowhead at each end. The line should extend from the top solid line to the next solid line. Next to it, on the left, write 'one cm.' Secondly, just below that, also on the right, draw another vertical arrowed broken line between the solid line and the horizontal broken line. Next to it, on the left, write 'one-and-a-half cm.' Thirdly, in the bottom right-hand corner, between the two solid lines, draw a vertical arrowed broken line. Next to it, on the left, write 'one cm.'

A Three-dimensional Figure

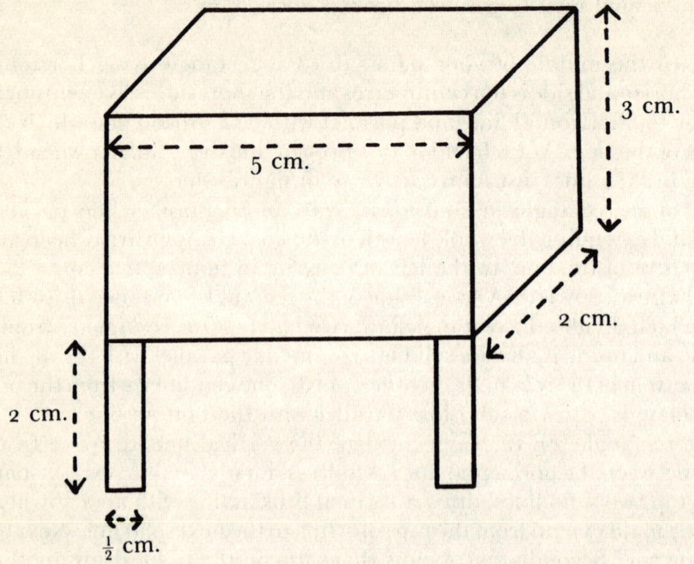

A Three-dimensional Figure

Introduction. The diagram is like a rectangular box, perhaps like a cupboard, with two legs in the front. Some of the dimensions are indicated by means of arrowed broken lines together with the measurements.

Description. Draw a rectangle in the centre of your paper which measures five centimetres along the horizontal side by three centimetres vertically. Just inside the rectangle, close to the top line, draw a horizontal broken line, with an arrowhead at each end, which extends the length of the line. Just under the line, and in the middle of it, write 'five cm.'

Listen carefully to the next part before drawing it. In perspective, draw the whole of the top of the box. It extends backwards, towards the right, at an angle of about forty-five degrees. The width is about two centimetres. In a similar way, draw the right-hand side of the box in perspective, joining it to the top and front. The top and side now look like two parallelograms.

On the right-hand side, next to the far right vertical line, and outside it, draw a vertical arrowed broken line. Next to it, on the right of it, and in the middle of it, write 'three cm.' Under the bottom of the right-hand side of the box, draw another arrowed broken line parallel with it, and next to it, in the middle of it on the right, put 'two cm.'

Listen to the following carefully. Looking at the front of the box, there are two short legs extending downwards. On the left draw two vertical parallel lines half-a-centimetre apart, extending down from the bottom of the box for about two centimetres. Join the two vertical lines at the bottom. To the left of the leg draw a vertical arrowed broken line the length of the leg, and on the left of it, in the middle, write 'two cm.' Under the bottom of the leg draw another small arrowed broken line, horizontally. Under it write 'half cm.'

Now draw the other leg to the right of the rectangle. It is exactly the same size as the first leg. However, do not draw any arrows or write any measurements on it.

An Abstract Design

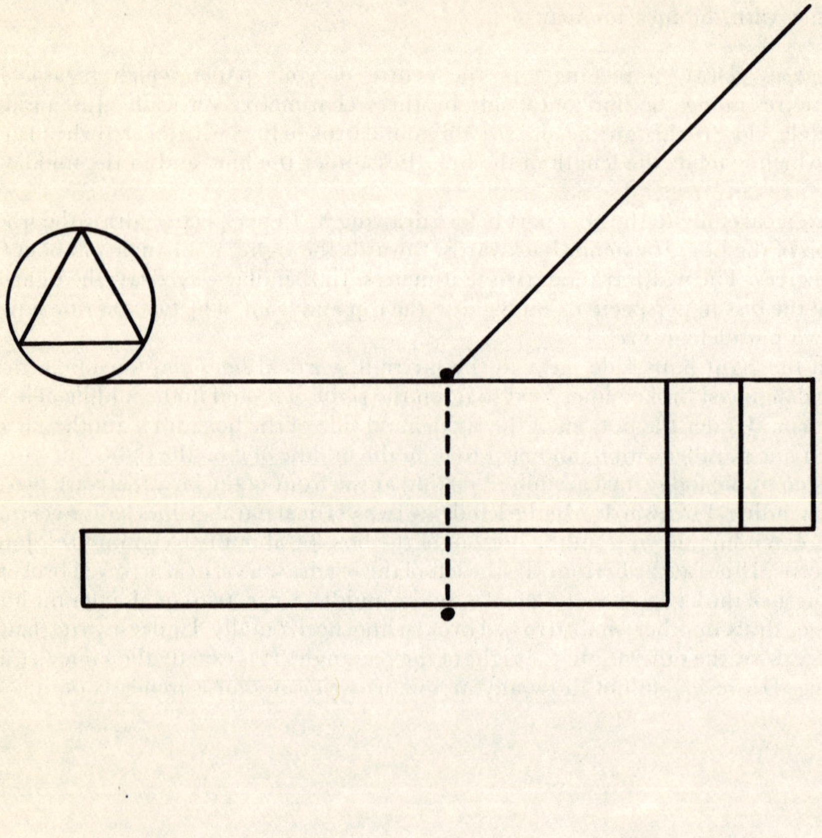

An Abstract Design

Introduction. This is an abstract design which includes straight lines, both horizontal and vertical, and solid and broken. It also includes rectangles, a square, a circle, a triangle, some large dots and a diagonal line.

Description. In the middle of your paper draw two horizontal lines, parallel with each other. They should be about ten centimetres long and about two centimetres apart. Join the two lines together at the extreme right of the lines by drawing a vertical line. Now form a square by drawing another vertical line to the left of the one you have just drawn. Continue the vertical line you have just drawn, downwards until it extends about one centimetre below the bottom horizontal line.

At the extreme left-hand end of the lower horizontal line draw a vertical line downwards so that it is the same length as the one at the right-hand end of the line. Now join the two vertical lines together by means of a horizontal line drawn parallel to the one above it.

Now look at the top horizontal line. Draw a big dot in the middle of it so that it is on top of the line. From the dot draw a diagonal line at about forty-five degrees to the right-hand side above the horizontal line. Continue drawing the line until the end of it is above the extreme right-hand end of the horizontal line.

Now, from the big dot draw a vertical broken line down to the bottom horizontal line. Where it touches the bottom line draw another big dot, just below the line.

Look now at the small square to the right of the horizontal lines. Draw a vertical line dividing the square into two halves.

Now look at the extreme left-hand end of the top horizontal line. On top of the line, at the very end, and just touching it, there is a small circle – about two centimetres in diameter. Now listen carefully. Inside the circle there is a triangle: all three points touch the sides of the circle. The base of the triangle is parallel to the horizontal lines. The apex of the triangle is above the centre of the base of the triangle.

A Geometric Figure

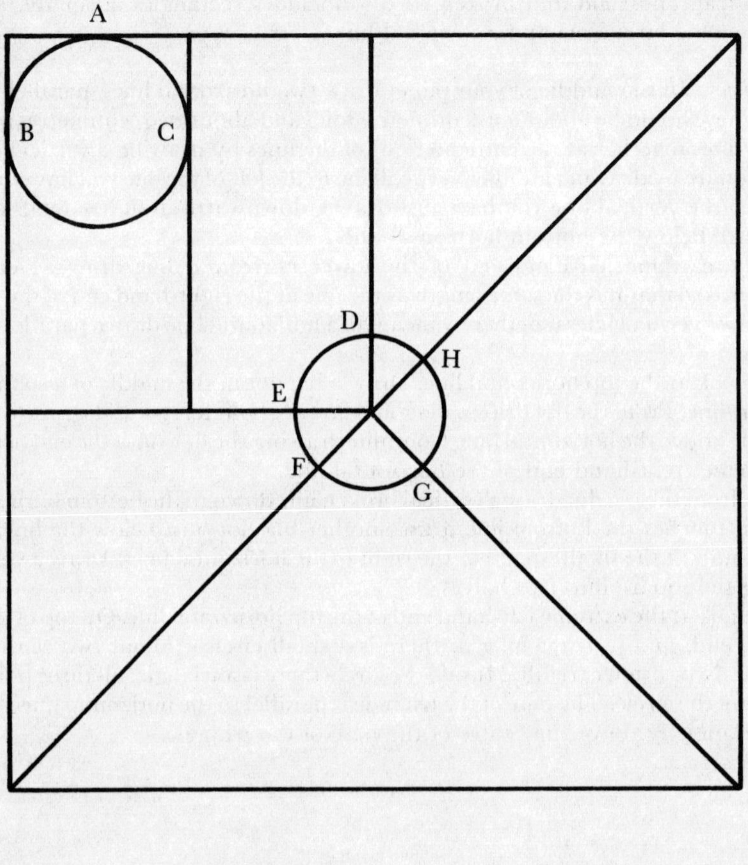

A Geometric Figure

Introduction. The diagram is a large square containing some diagonal, vertical and horizontal lines. It also contains two small circles, and the points at which they touch or cross other lines are labelled.

Description. In the centre of your paper draw a large square measuring about ten centimetres. Now draw a diagonal line from the bottom left-hand corner to the top right-hand corner. Draw another diagonal line from the bottom right-hand corner to the centre of the first diagonal line. It will meet it at right angles.

From the centre of the top side of the square draw a vertical line down which stops when it meets the first diagonal line in the centre. Then from the centre of the left-hand side of the square, draw a horizontal line to the centre of the square. It stops when it meets the other lines.

In the left-hand half of the large square you should now have a smaller square which measures about five centimetres. Halfway along the top side of the smaller square draw a vertical line down so that it divides the small square into two halves, in other words, into two rectangles. Inside the left-hand rectangle, draw a circle near the top so that it touches the top and the two sides of the rectangle. Then label the points where the circle touches the sides of the rectangle: at the top write a capital letter A outside the circle. To the left, but inside the circle, write the letter B. To the right, and inside the circle, write the letter C.

In the centre of the large square draw a slightly smaller circle than the first one. Draw it so that the centre of the circle is at the point where all the lines meet. You are now going to label with letters the points where the circle crosses the other lines. All the letters will be written outside the circle. Start at the top of the circle where it crosses the vertical line. Label this D above the circle and just to the left of the vertical line.

The other points will be labelled going round the circle to the left, in other words in an anti-clockwise direction. Label the next point E where the circle crosses the horizontal line. Write the letter E above the horizontal line. The next letter is F: write it just to the left of the diagonal line. The next letter is G: write it just to the left of the diagonal line which slopes down to the right. The final letter is H: write it just to the right of the top diagonal line.

Some Numbers

Table 1

108	118	180
880	818	3,116
3,066	3,806	3,006
3,860	3,160	3,816
3,060	3,106	103,116

Table 2

50	115	505
955	1,005	5,815
15,150	50,015	55,550
115,850	150,555	515,805

2 Data

Some Numbers

Introduction. There are two tables of numbers, one above the other. Each table has three columns and a number of rows. Every box in the tables contains a number.

Description. Listen to the following carefully before you begin to draw the two tables of numbers. Draw two fairly large rectangles, one above the other. The width of the rectangles is the same but the top rectangle should be longer than the bottom one. The measurements of the rectangles do not matter as long as the rectangles are big enough to contain all the numbers.

Divide both rectangles into three columns of equal width. Divide the top rectangle into five rows of equal size and the bottom one into four rows, also of equal size. Above the top rectangle, on the left, write Table one: the one is a number not a word. Above the bottom one, also on the left, write Table two. Underline both these headings.

Now write the following numbers in the appropriate boxes, writing the numbers from left to right across the columns. Start with Table one. In the first row, in the top left-hand corner, write a hundred and eight; next to it, a hundred and eighteen, and in the top right-hand corner, a hundred and eighty. In the second row write eight hundred and eighty; next, eight hundred and eighteen, and then three thousand, one hundred and sixteen. In the third row write three thousand and sixty-six; next, three thousand, eight hundred and six, and then three thousand and six. In the fourth row, write three thousand, eight hundred and sixty; next, three thousand, one hundred and sixty, and then three thousand, eight hundred and sixteen. In the bottom row, write three thousand and sixty; next, three thousand, one hundred and six, and then a hundred and three thousand, one hundred and sixteen.

Now turn to Table two. In the first row, in the top left-hand corner, write fifty; next to it, a hundred and fifteen, and in the top right-hand corner, five hundred and five. In the second row, write nine hundred and fifty-five; next, one thousand and five, and then five thousand, eight hundred and fifteen. In the third row, write fifteen thousand, one hundred and fifty; next, fifty thousand and fifteen, and then fifty-five thousand, five hundred and fifty. In the bottom row, write a hundred and fifteen thousand, eight hundred and fifty; next, a hundred and fifty thousand, five hundred and fifty-five, and then five hundred and fifteen thousand, eight hundred and five.

Two Families

(1) The Davies Family

Name	Date of Birth	Age at Death
Jean	1717	13
June	1770	80
Joan	1717	30

(2) The Davis Family

Name	Date of Birth	Age at Death
Jane	1880	18
John	1818	60
June	1880	40
Jean	1880	16
Joan	1818	19

Two Families

Introduction. There are two tables of information about two families, both with a surname that sounds the same but is spelled differently. The information given is the first or Christian name of the person, the year of birth, and the age at death.

Description. Listen to the following before you begin to draw the two tables of information. Draw two fairly large rectangles, one above the other: the top rectangle should be narrower and shorter than the bottom one. The measurements of the rectangles do not matter.

Divide both rectangles into three columns. Divide the top rectangle into four rows and the bottom one into six rows. In the first row of each rectangle write the following: in the top left-hand box write 'Name'; in the middle box write 'Date of Birth'; and in the top right-hand box write 'Age at Death'.

Above the first table, on the left, put the number one in brackets, and after it write 'The Davies Family': spell the surname Davies, D-a-v-i-e-s. Above the second table, also on the left, put the number two in brackets, and after it write 'The Davis Family': spell the surname, D-a-v-i-s.

In the first table, in the second row, under 'Name', write Jean, spelled J-e-a-n. Under 'Date of Birth' write, in figures, seventeen seventeen. In the third column, under 'Age at Death', write, in figures, thirteen. In the next row write June, spelled J-u-n-e; next to it put seventeen seventy; then put eighty. In the bottom row write Joan, spelled J-o-a-n; next to it put seventeen seventeen; and in the last box put thirty.

In the second table, in the row under 'Name' write Jane, spelled J-a-n-e. Under 'Date of Birth' write eighteen eighty. In the third column write eighteen. In the next row write John, spelled J-o-h-n; next to it put eighteen eighteen; then put sixty. In the next row write June; next to that put eighteen eighty; then put forty. In the following row write Jean; next to it put eighteen eighty; then put sixteen. In the bottom row write Joan; next to that put eighteen eighteen; and finally put nineteen.

U.K. Some Statistics

(in thousands)

	1966	1971	1975–76
Total marriages	437	459	431
Total divorces*	39.1	74.4	126.7
Live births	980	902	676
Deaths	643.8	645.1	662.5

* England and Wales only

U.K.: Some Statistics

Introduction. There is one quite large rectangular table giving some information about the United Kingdom in the form of social statistics for selected years.

Description. In the middle of your paper draw a rectangle measuring roughly ten centimetres horizontally by seven centimetres. Inside the rectangle, about one centimetre from the top, and parallel to it, draw a line across the rectangle. Now divide the rest of the rectangle in half by drawing another horizontal line across the rectangle. From the right-hand side of the rectangle measure inwards about two centimetres and draw a vertical line from the top to the bottom of the large rectangle. Measure inwards another two centimetres and draw another vertical line the length of the rectangle. Then draw one more vertical line another two centimetres from the previous one, again the whole length of the rectangle. You now have a large rectangular table, divided in half and which contains four columns, the left-hand one being roughly twice the width of the other three columns.

In the top narrow row of the table, leave the first box empty. In the next one, write in figures the year nineteen sixty-six. In the next box write nineteen seventy-one; and in the final one put nineteen seventy-five dash seventy-six. Just above the table, and on the right, put in brackets the words 'in thousands'.

In the top half of the table, just below the empty box at the top of the left-hand column, write the words 'Total marriages'. Just below that, and still in the top half of the table, write 'Total divorces'. At the end of the word 'divorces', and just above it, draw a small star or asterisk. Just under the table, on the left, draw another star or asterisk, and write immediately after it 'England and Wales only'. This is a qualifying footnote. In the bottom part of the table, at the top of the left-hand column, write 'Live births'. Just under that write 'Deaths'.

Now put the following figures in the columns under the different years. In the first row, opposite 'Total marriages' and under the year nineteen sixty-six, put four hundred and thirty-seven. In the next column put four hundred and fifty-nine. In the last column put four hundred and thirty-one. In the next row, opposite 'Total divorces', put the following figures: thirty-nine point one; seventy-four point four; a hundred and twenty-six point seven.

In the bottom half of the table put the following. Opposite 'Live births' put nine hundred and eighty, then nine hundred and two, and in the final column, six hundred and seventy-six. In the next row, opposite 'Deaths' put six hundred and forty-three point eight. In the next column put six hundred and forty-five point one; and in the final column put six hundred and sixty-two point five.

United Kingdom: Population

thousands

Year	United Kingdom	England and Wales	Scotland	Northern Ireland
1901	38,237	32,528	4,472	1,237
1911	42,082	36,070	4,761	1,251
1921	44,027	37,887	4,882	1,258
1931	46,038	39,952	4,843	1,243
1951	50,225	43,758	5,096	1,371
1961	52,709	46,105	5,179	1,425
1971	55,515	48,750	5,229	1,536
1981	55,671	49,011	5,117	1,543

United Kingdom: Population

Introduction. There is one large rectangular table giving population figures for selected years for the United Kingdom, and then separately for the countries that comprise the U.K., namely England and Wales, Scotland, and Northern Ireland.

Description. Draw a rectangle measuring approximately ten centimetres horizontally by nine centimetres. Inside the rectangle, about two centimetres from the top and parallel to it, draw a line across the rectangle. Now draw four vertical lines from top to bottom of the rectangle so that there are five equal columns.

In the box at the top of the left-hand column write the word 'Year'. In the next box write 'United Kingdom'; then 'England and Wales' in one box; then Scotland; and finally 'Northern Ireland'. Just above the table, and on the right, put the word 'thousands'.

In the left-hand column, under the word 'Year', put, in figures, the following eight years, one under the other: nineteen oh one, nineteen eleven, nineteen twenty-one, nineteen thirty-one, nineteen fifty-one, nineteen sixty-one, nineteen seventy-one, and finally, nineteen eighty-one.

In the next column, under 'United Kingdom', put the following figures so that they are opposite the years. At the top, opposite nineteen oh one, put thirty-eight thousand, two hundred and thirty-seven. The next figure down is forty-two thousand and eighty-two; then forty-four thousand and twenty-seven; then forty-six thousand and thirty-eight; then forty-four thousand and twenty-seven; then forty-six thousand and thirty-eight; then fifty thousand, two hundred and twenty-five; then fifty-two thousand seven hundred and nine; then fifty-five thousand, five hundred and fifteen and finally, fifty-five thousand, six hundred and seventy-one.

In the next column, under 'England and Wales', put thirty-two thousand, five hundred and twenty-eight at the top. Under it put thirty-six thousand and seventy; then thirty-seven thousand, eight hundred and eighty-seven; then thirty-nine thousand, nine hundred and fifty-two; then forty-three thousand, seven hundred and fifty-eight; forty-six thousand, one hundred and five; then forty-eight thousand, seven hundred and fifty; and finally, forty-nine thousand and eleven.

In the next column, under 'Scotland', put four thousand, four hundred and seventy two at the top; then four thousand, seven hundred and sixty-one; then four thousand, eight hundred and eighty-two; then four thousand, eight hundred and forty-three; then five thousand and ninety-six; then five thousand, one hundred and seventy-nine; then five thousand, two hundred and twenty-nine; and finally, five thousand, one hundred and seventeen.

In the end column, under 'Northern Ireland', put one thousand, two hundred and thirty-seven; then one thousand, two hundred and fifty-one; then one thousand, two hundred and fifty-eight; then one thousand, two hundred and forty-three; then one thousand, three hundred and seventy-one; then one thousand, four hundred and twenty-five; then one thousand, five hundred and thirty-six; and finally, one thousand, five hundred and forty-three.

NOTE: The population figures are taken from the official Census. The first Census in the U.K. was in 1801. It has been taken every ten years since then, with the exception of 1941 (because of the Second World War).

G.B. Road Accidents: Number Killed

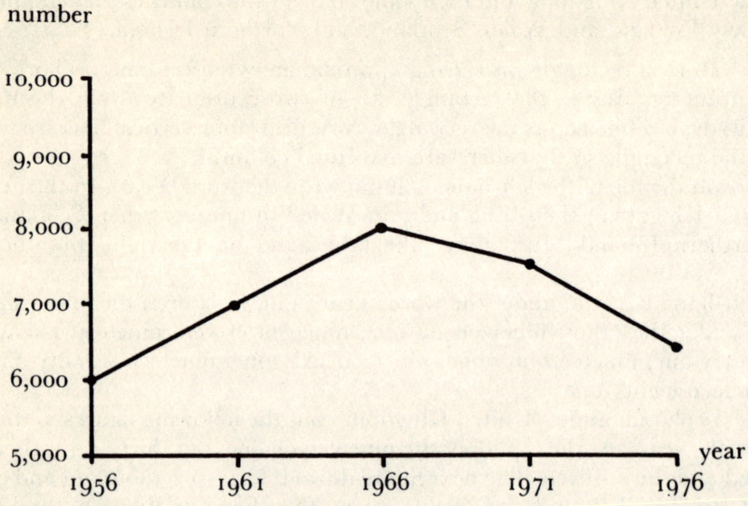

G.B. Road Accidents: Number Killed

Introduction. This is a graph containing one curve which shows the number of people killed in road accidents in particular years. The numbers are indicated on the vertical axis and the years on the horizontal axis.

Description. About three centimetres from the left-hand edge of your paper draw a vertical line about five centimetres long. Above it, and slightly to the left of it, write the word 'number'. This is the vertical axis. Now draw a horizontal line about eight centimetres long joined at the bottom of the vertical line and extending to the right. Just to the right of the horizontal line write the word 'year'. This is the horizontal axis.

Divide the horizontal axis into four equal parts by means of very small vertical marks. There are a total of five of the marks: one at each end of the horizontal line and three in between at equal distances from each other. Just below each mark write the years, in figures, which are given at five-yearly intervals. On the extreme left put nineteen fifty-six; next to it put nineteen sixty-one; then nineteen sixty-six; then nineteen seventy-one; and, finally, on the extreme right, nineteen seventy-six.

Now look at the vertical axis. Each centimetre put a small horizontal mark, and put one at the bottom as well; altogether, including the one at the very bottom, there will be six marks. To the left of each mark put a number as follows. At the bottom, level with the horizontal axis, put, in figures, five thousand. Next to the mark above that put six thousand. Above that put seven thousand, then eight thousand, then nine thousand, and finally, at the top, ten thousand.

On the graph plot the following positions and put a big dot for each one. Start with nineteen fifty-six: on the vertical axis put a dot by six thousand. Above nineteen sixty-one put a dot for seven thousand. For nineteen sixty-six, eight thousand. For nineteen seventy-one, seven thousand, five hundred; and for nineteen seventy-six, six thousand, five hundred.

When you have finished the dots, draw a line starting at the first dot for nineteen fifty-six and join up the dots by means of a straight line, finishing the line at the one for nineteen seventy-six.

The Trend to Monopoly in British Manufacturing

(Share of top hundred companies in Manufacturing output)

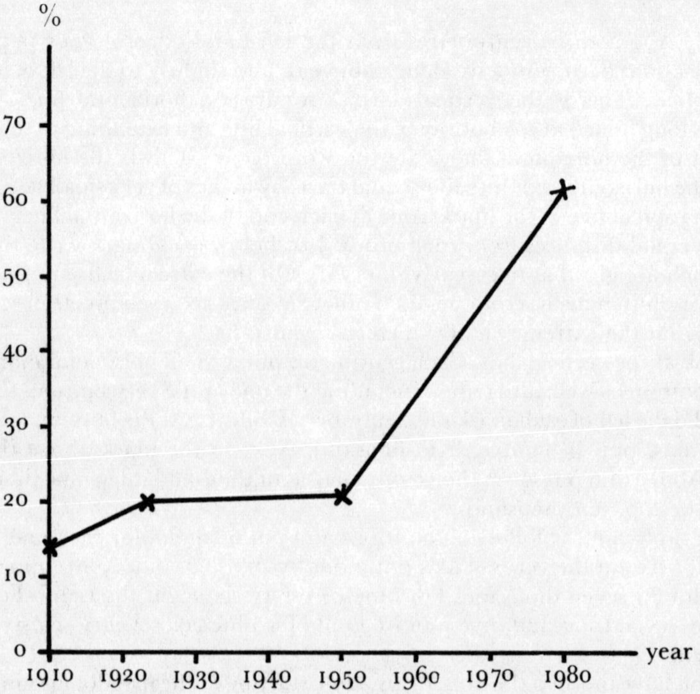

The Trend to Monopoly in British Manufacturing

Introduction. This is a graph containing one curve which shows the percentage share of the top companies in manufacturing output in particular years. The percentage share is indicated on the vertical axis and the years on the horizontal axis.

Description. About five centimetres from the left-hand edge of your paper draw a vertical line roughly eight centimetres long. Above it put the symbol for percentage (%). On this vertical axis mark off divisions at intervals of about one centimetre. At the very bottom, to the left of the mark at the base, put a zero. To the left of each mark going up the axis put the numbers, in tens, from ten to seventy.

Now draw the horizontal axis of the graph, also about eight centimetres long. Just to the right of the horizontal line write the word 'year'. Mark this line off in seven divisions at intervals of about one centimetre. On the extreme left, under the vertical axis, put in figures the year nineteen ten. Under the next mark put nineteen twenty. Then under the rest of the marks put the dates every ten years until the last one, which is nineteen eighty.

On the graph plot the following positions and put a very small cross for each one. Start with nineteen ten: on the vertical axis put a small cross by fourteen per cent. The next position is nineteen twenty-four with twenty per cent. After this comes nineteen fifty with twenty-one per cent. Finally comes nineteen eighty with sixty-one per cent.

When you have finished plotting the positions, draw a straight line joining up the crosses from the first one for nineteen ten to the last one at nineteen eighty.

The final thing to do is to write in the following subtitle in brackets under the main title: 'Share of top hundred companies in Manufacturing output.'

United Kingdom Energy Consumption

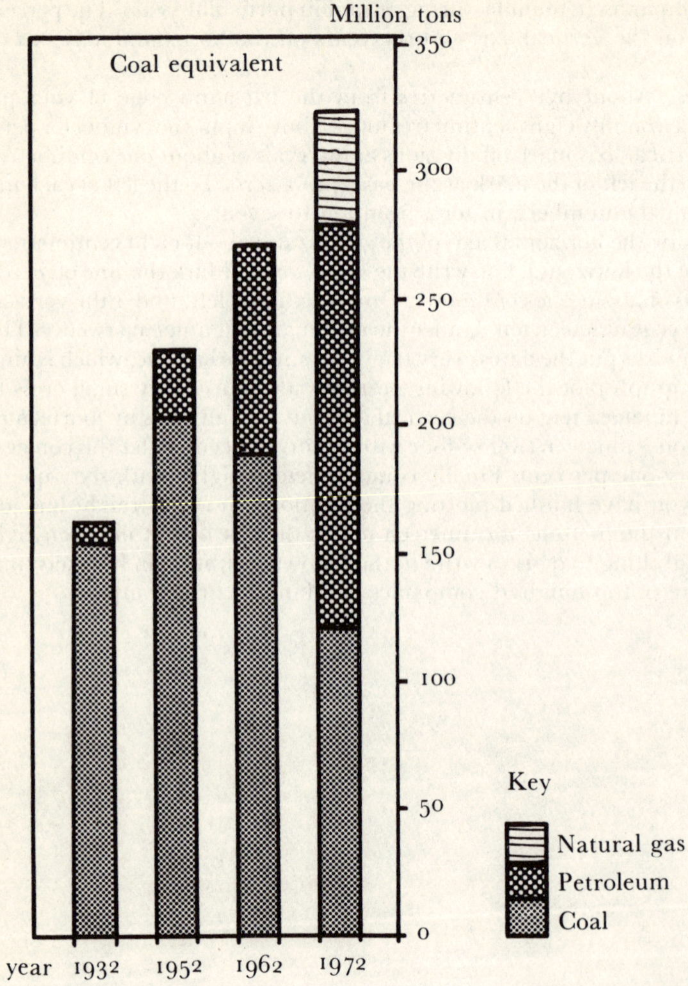

United Kingdom Energy Consumption

Introduction. This is a histogram, also called a bar diagram. It shows the consumption, in millions of tons, of natural gas, petroleum, and coal. The information is displayed in the form of vertical columns which are shaded in different ways for the different types of energy. The data is compared for four separate years.

Description. In the middle of your paper draw a rectangle standing vertically and measuring roughly five centimetres by twelve centimetres. Just above the top right-hand corner of the rectangle write the words 'Million tons'. This refers to the vertical axis. Divide the right-hand side of the rectangle into seven equal divisions, indicating each division by a very small mark on the line. Each division will be just over one and a half centimetres long. To the right of each division there are numbers which increase in units of fifty. To the right of the bottom right-hand corner put a zero. At the next mark above that put fifty, then a hundred, and so on to the top, which is three hundred and fifty.

To the left of the bottom left-hand corner write the word 'year'; this refers to the horizontal axis. Now listen carefully before drawing anything. The rectangle contains four vertical columns of different heights which are placed equi-distant along the baseline. The base of each column is labelled with a particular year. Now mark off the positions of the four columns along the horizontal axis, each one about half-a-centimetre wide, with roughly half-a-centimetre of space between each one. The first one starts about half-a-centimetre from the left-hand corner. Under the base of the one on the left put, in figures, the year nineteen thirty-two; under the next one put nineteen fifty-two; under the next nineteen sixty-two; and under the one on the right put nineteen seventy-two.

Now draw the heights of the columns as follows. The one for nineteen thirty-two extends up to between a hundred and sixty and a hundred and sixty-five million on the right-hand scale. When you have done this complete the column by drawing a short horizontal line across its top. The column for nineteen fifty-two goes up to about two hundred and thirty. The one for nineteen sixty-two goes up to about two hundred and seventy. The last one, for nineteen seventy-two, extends up to nearly three hundred and twenty-five. In a short while the content of the columns will be described.

But first listen to the following carefully before drawing anything. Just to the right of the histogram, at the bottom, there is a very small vertical rectangular box divided horizontally into three equal parts. Each part contains a pattern or form of shading to represent the different forms of energy. This shading is then used in the columns in the large rectangle. The very small rectangle is called a key. Now draw the small vertical rectangular box.

Above the very small rectangle write the word 'Key'. After you have divided the box into three equal parts horizontally, in the top part draw some parallel horizontal lines. To the right of them, outside the box, write the words 'Natural gas'. In the next part draw parallel diagonal lines in both directions so that they cross each other; to the right of them write the word 'Petroleum'. In the bottom part put a number of dots, to the right put the word 'Coal'. The Key is now complete.

Turn to the foot of page 49.

A Simple Machine

A PULLEY

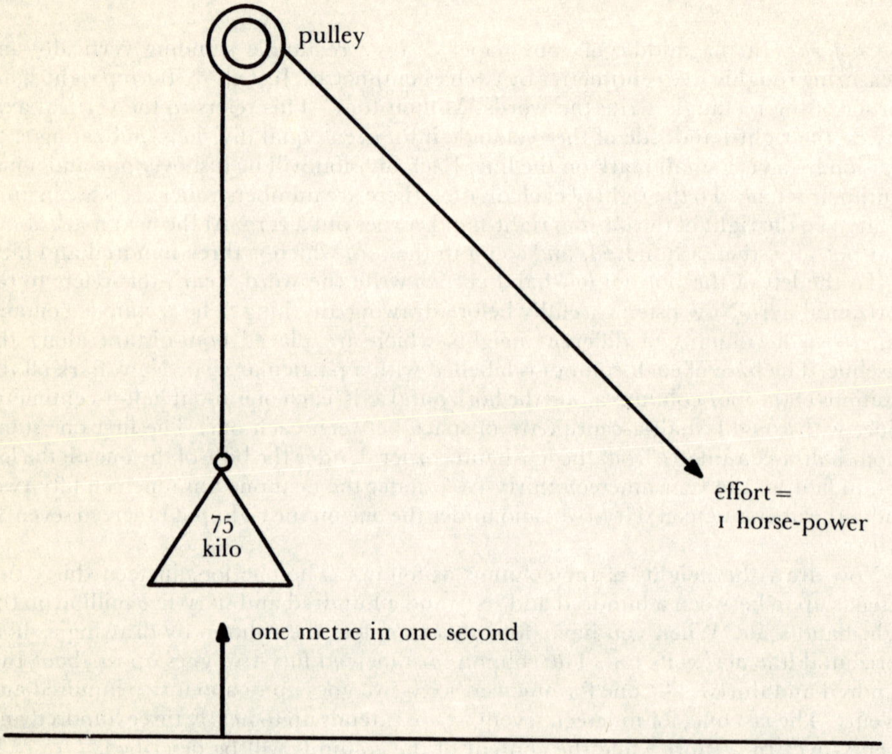

A Simple Machine – A Pulley

Introduction. This is a diagram of a simple machine, a pulley. It shows a weight being lifted from the ground by means of pulling a cord or rope which is passed over a small wheel, which is the pulley.

Description. The exact measurements of what you draw here are not important. In the centre of your paper draw a small circle; inside it draw a smaller circle. To the right of this wheel write the word 'pulley'.

From the left-hand side of the wheel draw a vertical line downwards, perhaps about five centimetres. This line represents the cord or rope. At the end of it draw a very small ring from which is suspended a triangle. Inside the triangle write in figures 'seventy-five kilo'. This represents the weight.

Now listen carefully before drawing anything else. From the right-hand side of the pulley wheel draw a line down to the right at an angle of about forty-five degrees. Stop the line when it is opposite the top of the triangle. At the end of the line draw an arrowhead which points downwards. Below and to the right of the arrowhead write the word 'effort' and then after it put an 'equals' sign. Starting under the word 'effort' put the number one and after it the words 'horse-power'.

About two centimetres below the triangular weight draw a horizontal line across the paper. Now listen carefully. From this line draw an arrow pointing up towards the centre of the baseline of the triangle. The arrow should extend about three-quarters of the distance from the horizontal line towards the weight. To the right of the arrow write 'one metre in one second'.

Contd. from page 47.

Now look again at the histogram. Look at the nineteen thirty-two column. Just below the top of the column, opposite a hundred and fifty-five, draw a horizontal line. In the small top part of the column draw crossed diagonal lines to represent petroleum, as in the Key. In the rest of the column put a number of dots to represent coal.

Turn now to the nineteen fifty-two column. Divide off the top of the column with a horizontal line opposite two hundred and five. Put in the lines to represent petroleum in the top part and dots in the lower part to represent coal.

The nineteen sixty-two column is done in a similar way. Draw a horizontal line opposite a hundred and ninety. Draw the top part to represent petroleum and the bottom part to represent coal.

The nineteen seventy-two column is divided into three parts. The top part is divided at two hundred and eighty. Draw parallel horizontal lines in the top part to represent natural gas. The next part is divided at a hundred and twenty. Put in crossed diagonal lines in the centre part to represent petroleum, and dots in the bottom part to represent coal.

Finally, write in the words 'Coal equivalent' just inside the top of the rectangle.

A Simple Machine

A LEVER

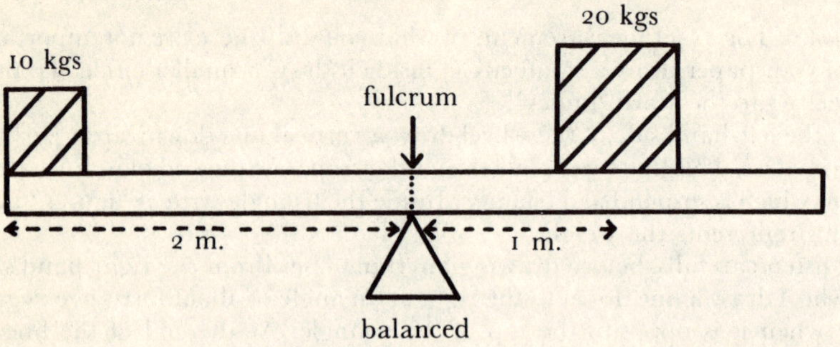

A Simple Machine – A Lever

Introduction. This is a diagram of a simple machine, a lever. It shows a long piece of wood balanced in the centre on a point. At one end of the wood is a square weight. It is balanced by another square weight, twice its weight, halfway along the other end of the wood.

Description. The exact measurements of what you draw are not important. In the centre of your paper draw a long, very narrow rectangle, lying horizontal: perhaps about ten centimetres by half-a-centimetre. This represents a long piece of wood. At the centre of the wood draw a short vertical dotted line across the wood so that it is divided into two halves. Just above the dotted line draw a small arrow pointing down at the dotted line, and above it write the word 'fulcrum', spelled f-u-l-c-r-u-m. Just below the dotted line, under the piece of wood, draw a small equilateral triangle so that the wood is balanced on the apex of the triangle. Under the triangle write the word 'balanced'.

On top of the piece of wood, at the left-hand end, draw a small square. Inside it draw some parallel diagonal lines going up from left to right. Above the square write in figures 'ten kgs', meaning ten kilograms. This represents a weight.

Listen carefully to the following. Halfway between the fulcrum and the right-hand end of the wood draw another, bigger square on top of the wood. Draw lines inside it in the same way as in the other square. Above the square write in figures 'twenty kgs'. This represents another weight.

Underneath the piece of wood, between the fulcrum and the left-hand end, draw a horizontal broken line with an arrowhead at each end pointing outwards. Under this broken line, and in the middle of it, put the figure 'two' and the letter 'm'. This indicates two metres in length. Similarly, on the right-hand side, under the wood, draw a broken line between the fulcrum and the centre of the large square. Draw arrowheads at each end of the line. Under the line, and in the centre of it, put the figure 'one' and the letter 'm'; this indicates one metre.

FACTORS AFFECTING HEALTH IN A DEVELOPING COUNTRY

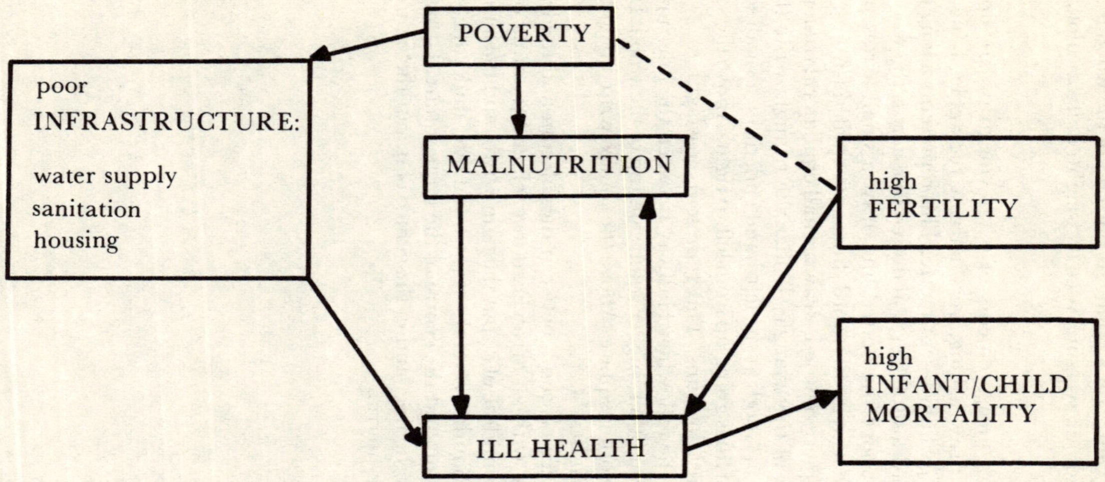

Factors Affecting Health in a Developing Country

Introduction. This is an organisational diagram showing the relationship between the different factors that can affect health in a developing country. The information is given in boxes that are linked by means of lines to indicate the direction of the relationship.

Description. The measurements of what you draw are not important in this diagram. In the centre of your paper draw a narrow rectangular box, lying horizontal, not more than about four centimetres long. Inside the box, in capital letters, write the word 'MALNUTRITION'. A short distance above this box, perhaps about one centimetre, draw another narrow rectangular box, slightly smaller. Inside this box, in capital letters, write the word 'POVERTY'. From the bottom edge of this box, in the centre, draw an arrow pointing downwards that touches the top of the 'Malnutrition' box.

About three centimetres below the 'Malnutrition' box, and the same size as it, draw another rectangular box. Inside the box, in capital letters, write the words 'ILL HEALTH'. From the bottom of the 'Malnutrition' box, on the left, draw an arrow pointing downwards that touches the top of the 'Ill Health' box also on the left. Draw another arrow from the top of the 'Ill Health' box, on the right, pointing up towards and touching the bottom of the 'Malnutrition' box also on the right.

To the left of the 'Malnutrition' box, just over one centimetre from it, draw a fairly large rectangular box, perhaps measuring roughly four centimetres across by three down. There are five lines of writing inside the box. At the top, write in small letters the word 'poor'. Immediately under it, in capital letters, write the word 'INFRASTRUCTURE' followed by a colon. There are three lines of writing under this, all in small letters. First write the words 'water supply'; then, underneath, write 'sanitation'; finally, immediately under that write 'housing'.

Now link this large box to 'Poverty' and 'Ill Health' as follows. Firstly, draw an arrow from the centre of the left side of the 'Poverty' box so that it points to and touches the top right-hand corner of the 'Infrastructure' box. Secondly, draw an arrow from the bottom right-hand corner of the 'Infrastructure' box down to the centre of the left side of the 'Ill Health' box.

On the right-hand side of the diagram there are two more rectangular boxes. The first one is to the right of 'Malnutrition' and nearly two centimetres from it. It is about the same width across as the 'Malnutrition' box but about twice its length downwards. Inside the box, at the top, write in small letters the word 'high', and just below it in capital letters the word 'FERTILITY'. Link this box to 'Poverty' by means of a broken line (not an arrow) from the centre of the left side of 'Fertility' to the centre of the right side of 'Poverty'. Link it also to 'Ill Health' by means of an arrow from the centre of the left side of 'Fertility' down to the top right-hand corner of the 'Ill Health' box.

The second box is about one centimetre below 'Fertility' and is the same width but slightly longer. Inside the box there are three lines of writing. At the top, in small letters, there is the word 'high'. Below it in capital letters write 'INFANT' stroke 'CHILD'. Under this is the final word, also in capitals 'MORTALITY'. Link this box to 'Ill Health' by means of an arrow from the centre of the right side of 'Ill Health' pointing up to the centre of the left side of 'Mortality'.

A Timetable

day \ time	9.30–10.30	10.30–11.30	11.30–12.30	2–3	3–4
MONDAY	lecture – main hall	seminar – Room 19	free	seminar – Room 70	library
TUESDAY	lecture – small hall	lecture – Room 213	tutorial – Mr Johns	language class	library
WEDNESDAY	see supervisor	meeting in Room 312	seminar – Room 117	sports	sports
THURSDAY	lecture main hall	lecture – Room 130	Mr Jones' class	language class	library
FRIDAY	group project	group project	discussion group	free	free

A Timetable

Introduction. This is a timetable for a student in an educational institution such as a college or a university. It shows typical morning and afternoon periods that a student may have during the week.

Description. The measurements of the following are not important. On your paper draw a very large rectangle with the horizontal side slightly longer than the vertical side. Divide the rectangle vertically into six equal columns. Inside the rectangle draw one line parallel with the top horizontal line and about one centimetre from it so that a narrow box is formed across the top of the rectangle. Next, divide the rest of the rectangle into five equal rows.

In the small box in the top left-hand corner of the rectangle draw a diagonal line from the top left-hand corner down to the bottom right-hand corner of the box. In the right-hand half of the small box write in small letters the word 'time', and in the left-hand half write the word 'day'. Write the following times, in figures, in the five small boxes at the top of the columns, from left to right. In the first one put nine thirty dash ten thirty. In the second one put ten thirty dash eleven thirty. In the next one put eleven thirty dash twelve thirty. In the fourth one put two dash three; and in the last one put three dash four.

In the extreme left-hand column, in the five boxes under the word 'day', write in capital letters the five weekdays beginning with MONDAY . . . TUESDAY . . . WEDNESDAY . . . THURSDAY . . . FRIDAY.

In the top row, for Monday, from left to right, fill in the five periods or sessions as follows. First, lecture dash main hall. Next, seminar dash Room nineteen. The next period is free. The afternoon periods are: seminar dash Room seventy; lastly, library.

Tuesday's periods are as follows: lecture small hall; then lecture Room two hundred and thirteen; then tutorial Mr. Johns. The afternoon periods are: language class followed by library. For Wednesday put first see supervisor; then meeting in Room three hundred and twelve; and seminar Room one hundred and seventeen. In the afternoon there is a double sports period.

Thursday's periods start with lecture main hall; followed by lecture Room one hundred and thirty; and then Mr. Jones' class. In the afternoon there is a language class followed by a library period. The first two periods on Friday morning are for a group project, followed by a discussion group. The afternoon is free.

A Page from a Diary

OCTOBER

12 Sunday 7.15 pm meet Ann outside the ABC Cinema	**16 Thursday** 10 am book theatre tickets 7.30 pm dinner with Mr & Mrs Green
13 Monday a.m. return books to the library 4.10 pm Students' Union meeting	**17 Friday** 8 pm theatre
14 Tuesday 11.20 am interview with Mr Johns	**18 Saturday** 2.30 pm football match 8.45 pm Ben's party
15 Wednesday 1.45 pm International Society coach trip	**Notes** Finish essay for next Monday

A Page from a Diary

Introduction. This shows the page from a diary for one week in the month of October. The dates and days of the week are given together with some entries about activities and times. The information is shown in two columns.

Description. The measurements of the following are not important. On your paper draw a very large rectangle with the horizontal side slightly longer than the vertical side. Divide the rectangle in half by means of a vertical line. Now listen carefully to the following. Divide the left-hand half into four boxes by means of three horizontal lines as follows: the top box is slightly smaller than the other three which are all the same size. Draw the bottom horizontal line so that it extends right across the right-hand half of the rectangle as well.

The bottom box is already provided in the right-hand half. Listen carefully to the following before drawing. Divide the rest of the right-hand half into three boxes by means of two horizontal lines: the third box down is slightly smaller than the two above it; they are the same size as the fourth box.

You have now drawn an empty diary page. Just above it, in the centre, in capital letters write the word 'OCTOBER'. Each box, except the bottom right-hand one, represents a day of the week. Put the dates that follow in figures and the days in small letters, except for the first capital letter.

In the top left-hand box, in the top left-hand corner, write 'twelve' followed by 'Sunday'. In the next box down put 'thirteen Monday'. In the next one down put 'fourteen Tuesday' and in the bottom one on the left put 'fifteen Wednesday' In the top right-hand box, on the left, put 'sixteen Thursday'; under it put 'seventeen Friday' and then 'eighteen Saturday' In the bottom right-hand box, on the left, put the heading 'Notes' and underline it.

To return to the first box. Under Sunday put 'seven fifteen pm' and next to that write 'meet Ann outside the ABC cinema'. Under Monday put 'a.m.' and next to it write 'return books to the library'. In the same box, and under a.m. put 'four ten pm' and next to it write 'Students' Union meeting'. Under Tuesday put 'eleven twenty am' and next to it 'interview with Mr. Johns'. Under Wednesday put 'one forty-five pm' and next to it write 'International Society coach trip'.

Under Thursday put 'ten am' and next to it write 'book theatre tickets'. Also for Thursday put 'seven thirty pm dinner with Mr and Mrs Green'. For Friday put 'eight pm theatre'. For Saturday put 'two thirty pm football match'. Just under that put 'eight forty-five pm Ben's party'.

Finally, in the box headed 'Notes' write 'Finish essay for next Monday'.

An Island

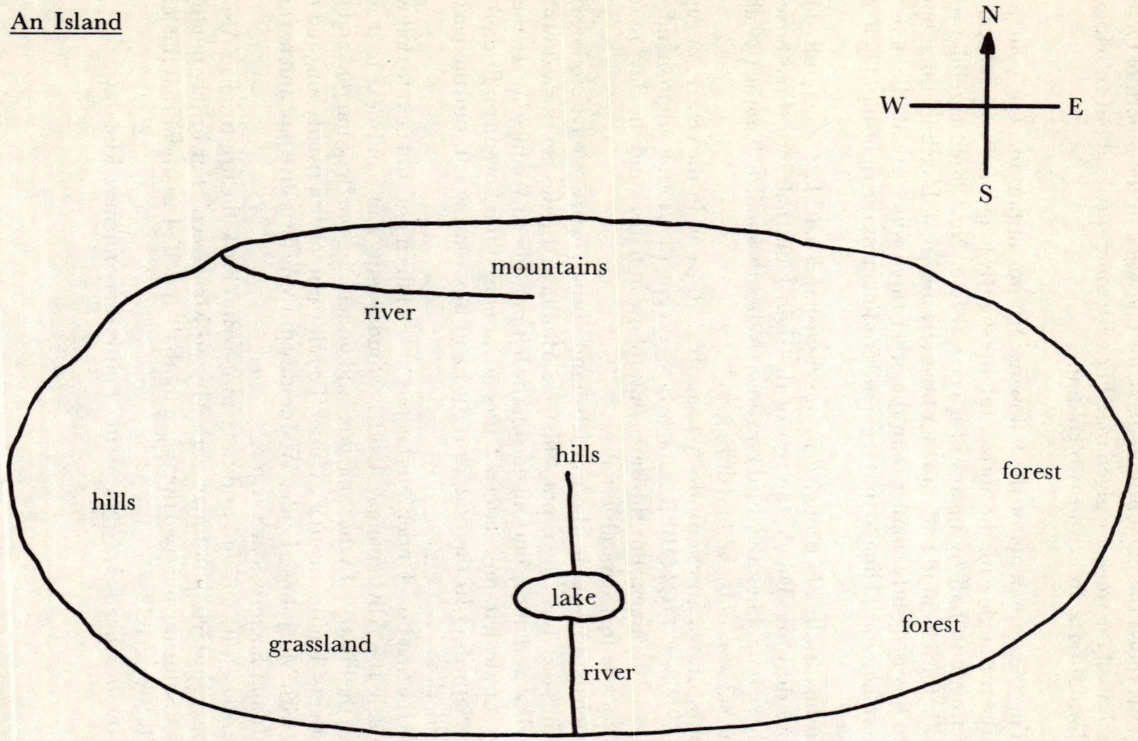

scale: 1 cm. to 1 km.

3 Maps and Plans

An Island

Introduction. This is a drawing of an island which is oval in shape, or egg-shaped. It contains a number of land features which are indicated by words not pictures.

Description. In the middle of your paper draw a large oval, measuring roughly fifteen centimetres across, by about seven centimetres (or half its width) down. When you have done that, in the top right-hand corner of the paper draw an arrow pointing upwards and above it write the letter N in capitals. This indicates North. At the bottom of the arrow write the letter S. Across the centre of the arrow draw a horizontal line the same length as the arrow. At the end of the line on the left write the letter W. At the end on the right put the letter E.

Now look at the oval shape, the island. Write the words 'mountains', in small letters, in the north of the island. All the words that you will have to write in the island will be in small letters. In the extreme west of the island write the word 'hills'. In the extreme east of the island write the word 'forest'. Write the word 'forest' again in the south-east of the island. In the south-west, write the word 'grassland'. In the centre of the island write the word 'hills'. A short distance below the word 'hills' write the word 'lake and draw an oval or circle around it. From the word 'hills' down to the lake draw a straight line: this represents a river. Continue the river from the south side of the lake down to the south of the island. To the right of the river, and next to it, write the word 'river'. The final thing to draw in the island is a line from the word 'mountains' curving inland and then going to the north-west coast of the island. This represents another river. Write the word 'river' just below the line.

Finally, under the island, on the left, write the word 'scale' in small letters, followed by a colon and then, in figures, 'one cm. to one km.' Underline all of it. This indicates that one centimetre on the map is equal to one kilometre on the ground.

Country X: Railway Network

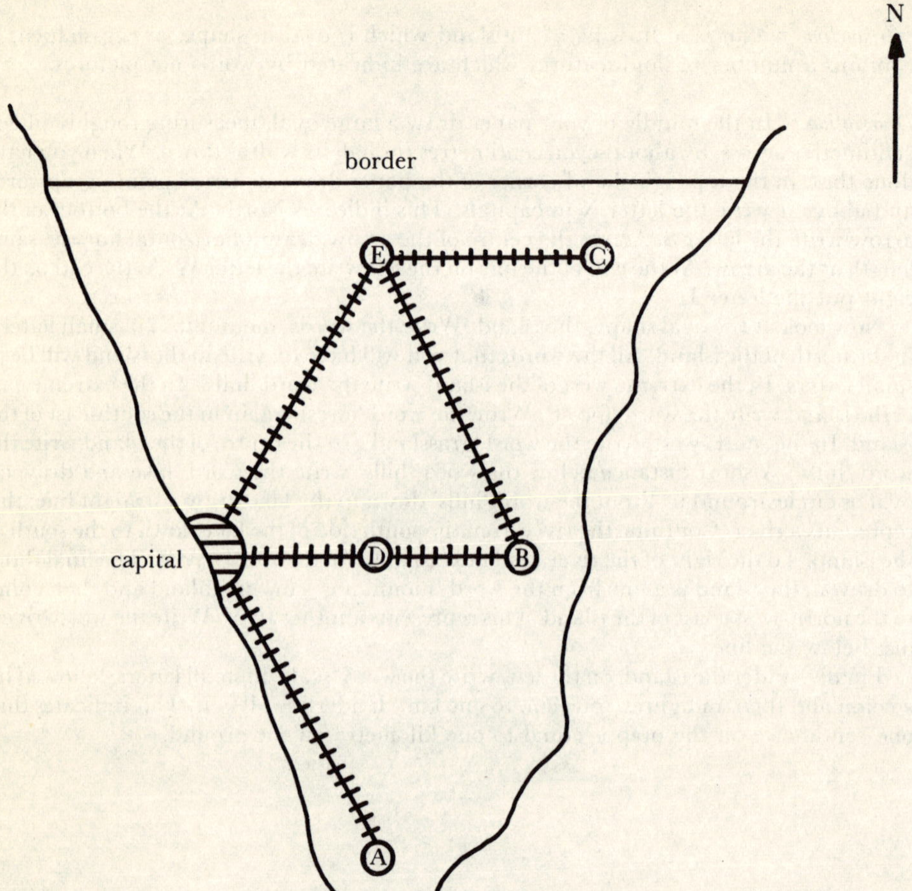

scale: 1:6 000 000

Country X: Railway Network

Introduction. This is a map of a country showing how the railway system is linked from city to city and to the capital city. The shape of the country is similar to India or the lower part of South America.

Description. Listen very carefully to the following before drawing it. On your paper draw the outline of the country in the shape of a very large letter V but draw the lines slightly wavy, and the bottom of the V should be curved and not pointed. The outline represents the coastline. Now draw the outline. Near the top of the letter V draw a straight horizontal line right across the top so that the two sides of the country are joined. Just above the centre of the straight line write the word 'border'. To the right of the top of the country draw a small arrow pointing upwards and above it write the capital letter N to represent North. Just below the bottom of the country write the word 'sea'. To the right of that write the following: 'scale' followed by a colon and the figures one colon six and six noughts. In other words, the scale is one to six million. This line is underlined.

Halfway down the west coast of the country draw a half circle which projects inland and measures roughly one centimetre in diameter. Shade it in with parallel horizontal lines and to the left of it write the word 'capital'. This represents the capital city of the country. Other cities in the country will be represented by small circles inside which are the capital letters, A, B, C, D and E.

Just below the centre of the border draw a small circle and inside it put the letter E. Parallel with that and near the east coast draw another circle and put the letter C inside it. Opposite the capital city and directly below city E draw another circle and label it D. Opposite D, so that it is in a straight line with D and the capital city, draw another circle near the coast. Label this B. The final city, A, is in the extreme south of the country in a direct line with E and D. It, too, is near, not on, the coast.

Now join the cities together with the railway line as follows. The railway line is represented by a straight line with very short lines crossing it at regular intervals. Join C to E, and E to the capital city. Then join E to B, B to D, and D to the capital. Finally, join A to the capital.

Front View (cross-section) of a House

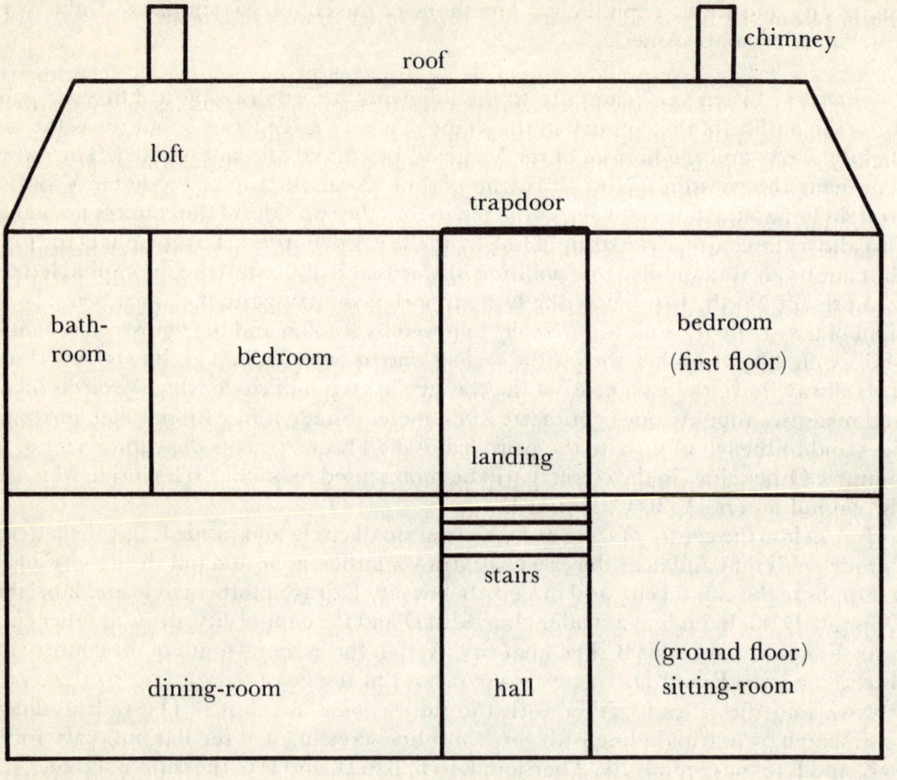

scale: 1:90

Front View (cross-section) of a House

Introduction. This is a diagram of a two-storey house, showing a cross-section of the building from the front. The rooms and some other items are labelled.

Description. Draw a rectangle, lying horizontally, measuring about twelve centimetres by seven. This represents the house. Now draw the roof on the house as follows: draw a line parallel with the top of the rectangle and nearly two centimetres above it. Stop the line about one centimetre from each end of the house. Now join the top corners of the house to the horizontal line by means of a sloping line at each end. This is the roof.

On top of the roof draw two small rectangles, one near each end, about one centimetre high and half a centimetre wide, to represent the chimneys. To the right of the right-hand chimney write the word 'chimney'. Just above the roof, slightly to the left of centre, write the word 'roof'. Inside the roof, just below the left-hand chimney, write the word 'loft'.

Now look at the house itself. Divide it in half horizontally by means of a straight line. Also divide it in half vertically by means of another straight line. Two centimetres to the right of the vertical line draw another vertical line of the same length. The house is now divided into rooms, both upstairs and downstairs. One more room must be added upstairs and then the rooms can be labelled. Two centimetres from the left-hand wall of the house draw a vertical line dividing the upstairs section only.

In the small upstairs room that you have just drawn on the left, write the word 'bathroom' in the centre. In the room next to it, also in the centre, write the word 'bedroom'. In the narrow part next to the bedroom, and near the bottom of it, write the word 'landing'. On the line immediately above the landing draw a very thick horizontal line the width of the landing. Just above it write the word 'trapdoor'.

In the centre of the room, to the right of the landing, write the word 'bedroom'. Just under this word put in brackets 'first floor'. You have now finished the upstairs or first floor of the house.

In the downstairs room on the left write 'dining-room' in the centre near the bottom. In the narrow part next to the dining-room write 'hall' near the bottom. At the top of the hall draw four parallel horizontal lines the width of the hall and quite close together. Just under these lines write the word 'stairs'. In the room to the right of the hall write 'sitting-room' near the bottom. Just above this word write, in brackets, 'ground floor'.

You have now finished drawing the house. The final thing to do is to write the word 'scale' followed by a colon just under the bottom left-hand corner of the house. The scale is one to ninety. Represent this by putting in figures one colon ninety immediately after 'scale' then underline it.

NOTE: In such diagrams the front view is often referred to as the *elevation.*

GROUND PLAN OF A HOUSE

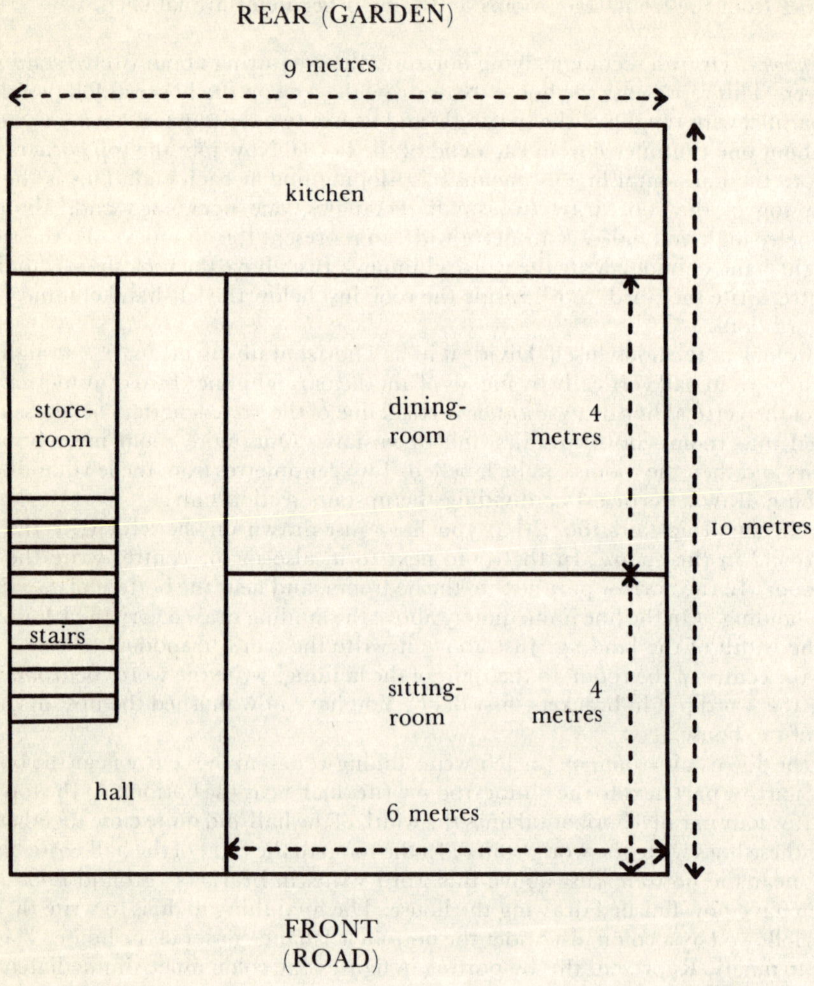

Ground Plan of a House

Introduction. This is a plan of a house from above showing the division of the ground floor into rooms. Some dimensions are given in metres and they are indicated by means of arrowed broken lines.

Description. In the middle of your paper draw a large rectangle measuring nine centimetres across by ten centimetres down. Just above the top of the rectangle, and parallel with the length of it, draw a horizontal broken line with arrowheads at each end. Just above this, and in the middle of it, write the number nine followed by the word 'metres'. Just above this, in capital letters, write the word 'REAR' – spelled R-E-A-R, and in brackets after it, also in capitals, the word 'GARDEN'.

To the right of the house, close to it and parallel with the length of the right-hand side, draw a vertical broken line with arrowheads at each end. Just to the right of this, and in the middle of it, write 'ten metres'. Just below the rectangle, and in the centre, write the word 'FRONT' in capital letters. Immediately below this word, also in capitals, and in brackets, write 'ROAD'.

Now look at the outline of the house. Inside the house and two centimetres from the top draw a horizontal line across the rectangle. In the centre of the narrow section that you have drawn write the word 'kitchen' in small letters. Inside the house and three centimetres from the left-hand wall of the house draw a vertical line from the kitchen down to the front wall of the house. Inside this narrow section that you have just drawn write the word 'hall' near the front of the house.

Inside the hall, starting from the wall with the kitchen, draw a vertical line dividing the hall into two halves but stop the line about two centimetres from the front wall of the house. Draw a horizontal line from the end of this dividing wall in the hall to the left-hand wall of the house. Now divide in half the left-hand narrow section that you have in the hall by means of a horizontal line. Above the line write the word 'storeroom'. Below the line draw a number of parallel horizontal lines to represent stairs, and between two of them write the word 'stairs'.

Now look at the remaining section of the house, the rectangle to the right of the hall. Divide this in half by means of a horizontal line. You have now created two smaller rectangles. In the upper one, in the centre of it, write the word 'dining-room'. In the lower one, also in the centre, write the word 'sitting-room'. Next to the right-hand wall of the dining-room and parallel with it, draw a vertical arrowed broken line the length of the dining-room wall. Just to the left of it, and in the middle of it, put 'four metres'. Do exactly the same in the sitting-room below. Also in the sitting-room draw a horizontal arrowed broken line next to, and parallel with, the front wall. Just above it, and in the middle of it, write 'six metres'.

A Town Centre: Sketch Map

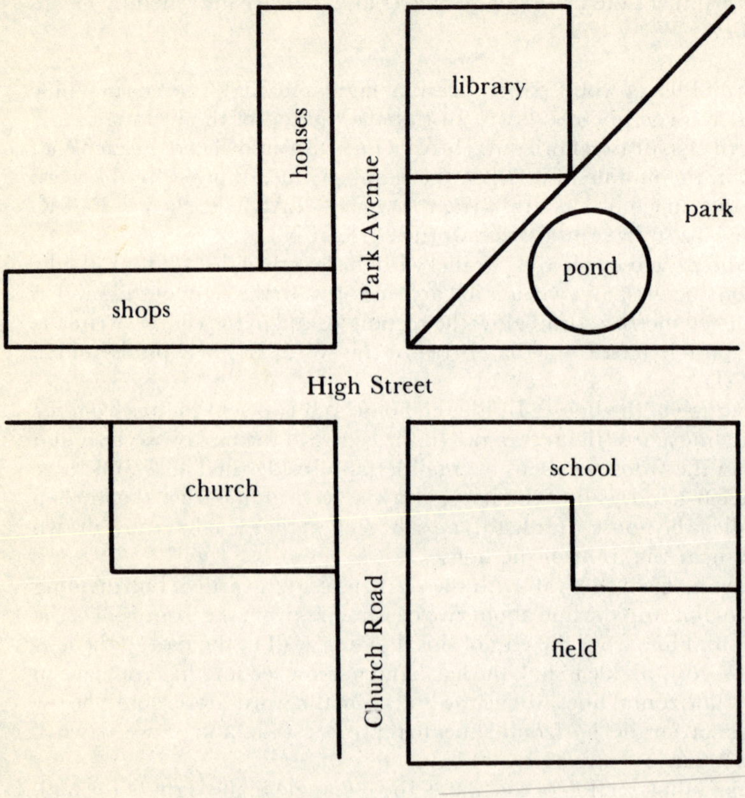

A Town Centre: Sketch Map

Introduction. This is an outline map of a town centre. It shows a crossroads with a number of buildings and other features marked on it. The measurements are not important as the map is not to scale.

Description. Listen carefully to this first part before drawing anything. In the middle of your paper draw a large crossroads as follows. There are two parallel vertical lines about one centimetre apart and about ten centimetres long; they are crossed in the centre by two similar horizontal lines. Between the horizontal lines, and in the centre of them, write the words 'High Street'. Between the upper vertical lines write 'Park Avenue'. Between the lower vertical lines write 'Church Road'.

On the left-hand corner of the High Street and Church Road draw a rectangle so that the long side is along the High Street and the short side is along Church Road. The dimensions do not matter. Inside the rectangle write the word 'church'.

In the section of the map to the right of Church Road draw in two more lines, one vertical and one horizontal, so that a square is formed with the whole of Church Road and the High Street. Inside this square draw a horizontal line from the middle of the right-hand side to the centre of the square. Then continue the line vertically upwards stopping a short distance from the High Street. Then continue the line horizontally to the left stopping it when it reaches Church Road. In the upper section that you have drawn write the word 'school' and in the section below the horizontal line write 'field'.

On the left-hand corner of the High Street and Park Avenue draw a narrow rectangle so that the long side is the whole length of the High Street and the very short side is along Park Avenue. In the centre of this rectangle write 'shops'. Draw a similar rectangle along the side of Park Avenue, from the shops to the end of Park Avenue. In the centre of this write 'houses'.

Listen carefully to the next part before drawing it. On the other corner of the High Street and Park Avenue draw a diagonal line at about forty-five degrees so that it divides the right-hand corner into two. Draw the diagonal line up to the right so that it stops opposite the top of Park Avenue. About half-way down Park Avenue, on the right, draw a horizontal line from the road to the centre of the diagonal line. Now draw another horizontal line parallel with it starting from the top of Park Avenue and stopping immediately above the end of the first one. Join the two lines together by means of a vertical line. Inside the square that you have drawn write 'library' in the centre.

The last part of the map is as follows. To the right of the diagonal line, about half-way along this end of the High Street, draw a circle that nearly touches both the diagonal line and the High Street. Inside the circle write 'pond'. Just to the right of the pond write 'park'.

A university: part of the campus

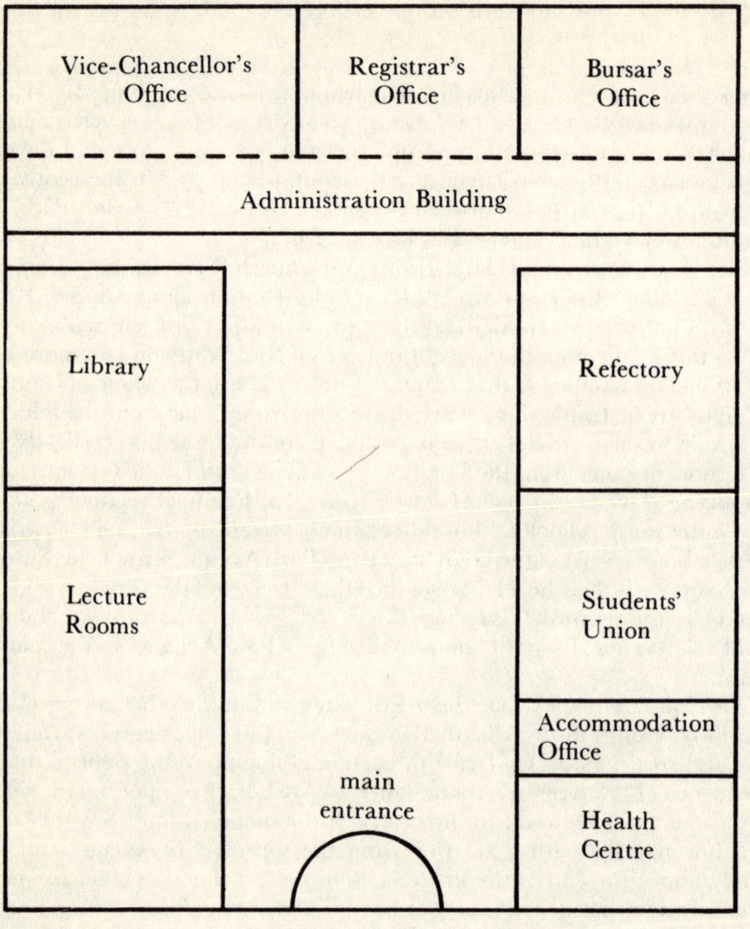

A University: Part of the Campus

Introduction. This is a plan of part of a university campus. It shows a number of buildings normally associated with a university. The plan is not drawn to scale.

Description. In the middle of your paper draw a rectangle standing vertically and measuring about ten centimetres by twelve. About three centimetres from the top of the rectangle, and inside it, draw a horizontal line which extends the whole width of the rectangle. About one centimetre above this line draw a broken line which is parallel to it and the same length. Between the broken line and the solid line beneath it write, in the middle, 'Administration Building'.

Divide the space above the broken line into three parts by means of vertical lines, making sure that the left-hand section is slightly wider than the other two which are the same width. These three sections represent three offices inside the Administration Building. Inside the left-hand one write 'Vice-Chancellor's Office': Vice-Chancellor is spelled V-i-c-e hyphen C-h-a-n-c-e-l-l-o-r. Inside the middle one write 'Registrar's Office': Registrar is spelled R-e-g-i-s-t-r-a-r. Inside the right-hand one write 'Bursar's Office': Bursar is spelled B-u-r-s-a-r.

Now look at the empty part of the rectangle below the Administration Building. On the left draw a column about three centimetres wide but leave a narrow space, perhaps half-a-centimetre, between it and the Administration Building. Draw a horizontal line across the top of the column. Opposite this, on the right-hand side of the rectangle, draw another column, the same width with a similar narrow space left between it and the Administration Building.

Divide the right-hand column into three equal parts by means of horizontal lines. Sub-divide the bottom part of the column by drawing another horizontal line about one centimetre from the top. You now have four boxes on the right. Inside the top one, in the middle, write 'Refectory'. Inside the next one write 'Students' Union'. In the small one under this write 'Accommodation Office'; and in the bottom one write 'Health Centre'.

Now look at the column on the left. About one-third of the way down it draw a horizontal line so that it is divided into two parts. Inside the upper part, in the middle, write 'Library'. Inside the lower part write 'Lecture Rooms'.

The plan is almost finished. The only thing to add now is the main entrance. This is done by drawing a small arc or semi-circle on the baseline of the rectangle between the Lecture Rooms and the Health Centre. Just above the arc write 'main entrance'.

Part 1 (B) Practice: Answers

The following answers are suggestions: some small variations may be possible, particularly in *Exercise 7*.

Exercise 1

H	B	E
k		G
A		m
	j	C
i	F	
D	l	

(1) B is in the middle column, top row.
E is in the top right-hand box.
G is in the right-hand column, second row down.
A is in the left-hand column, third row down.
C is in the right-hand column, fourth row down.
F is in the middle column, fifth row down.
D is in the bottom left-hand box.

(2) See the small letters in the table.

(3) Various answers are possible.

Exercise 2

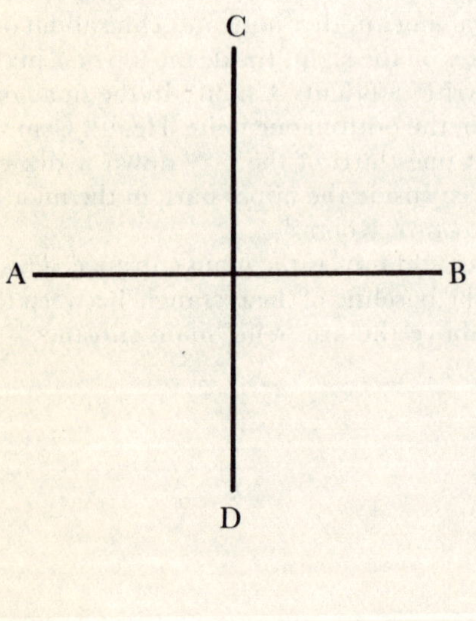

Exercise 3

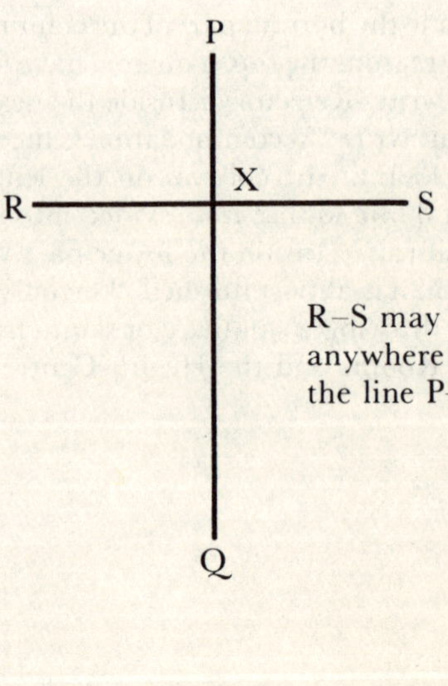

R–S may be anywhere along the line P–Q